Philosophies and Sociologies of Bioethics

Hauke Riesch • Nathan Emmerich
Steven Wainwright
Editors

Philosophies and Sociologies of Bioethics

Crossing the divides

 Springer

Editors
Hauke Riesch
Department of Sociology
and Communications
Brunel University
London, UK

Steven Wainwright
Department of Social & Political Sciences
Brunel University London
Uxbridge, Middlesex, UK

Nathan Emmerich
ANU Medical School
Australian National University
Canberra, Australia

The Institute of Ethics,
Dublin City University
Dublin, Ireland

School of History, Anthropology,
Politics and Philosophy
Queen's University Belfast
Belfast, UK

ISBN 978-3-319-92737-4   ISBN 978-3-319-92738-1   (eBook)
https://doi.org/10.1007/978-3-319-92738-1

Library of Congress Control Number: 2018948009

Printed on acid-free paper

This Springer imprint is published by the registered company Springer International Publishing AG part of Springer Nature.
The registered company address is: Gewerbestrasse 11, 6330 Cham, Switzerland

Contents

Introduction: Crossing the Divides 1
Hauke Riesch, Nathan Emmerich, and Steven Wainwright

Elective Modernism and the Politics of (Bio)Ethical Expertise 23
Nathan Emmerich

Grounding Knowledge and Normative Valuation in Agent-Based
Action and Scientific Commitment 41
Catherine Kendig

Ethics and Citizen Participation in the uBiome Institutional
Review Board Debate: Some Reflections on Social
and Normative Analyses 65
Lorenzo Del Savio

Minding the Gaps: Sensitivities in Pursuing Empirical Ethics 77
Brian Rappert

'It's Not Just About Having Babies': A Socio-bioethical Exploration
of Older Women's Experiences of Making Oncofertility
Decisions in Britain 93
Alexis Paton

'Can Someone Please Decide?' How the Media Represent the Risk
of Drinking During Pregnancy 107
Hauke Riesch

The Ethical Framework for the Use of E-Cigarettes 127
Nancy Tamimi

Performing Risk & Ethics in Clinicians' Accounts of Stem Cell
Liver Therapies 149
Steven Wainwright, Mike Michael, and Clare Williams

Outroduction 171
Hauke Riesch, Nathan Emmerich, and Steven Wainwright

Contributors

Lorenzo Del Savio earned his PhD at the European School of Molecular Medicine, Milan. He took part in a research project on the social, ethical and regulatory aspects of citizen science in biomedicine at the University of Kiel.

Nathan Emmerich is a research fellow in the Institute of Ethics, Dublin City University, and a visiting research fellow in the School of History, Anthropology, Politics and Philosophy at Queen's University Belfast. His research falls under the broad rubric of Bioethics. Whilst his background is in philosophy and 'applied ethics', his interests are interdisciplinary. He is less interested in applied ethics and normative solutions to the dilemmas of medicine and medical practice than with 'meta-bioethical' questions of ethics as an aspect of professional practice and our broader culture and society. To this end he is mostly interested in using social theory to examine how morality and ethics are produced, reproduced and 'done' in various domains of medicine and healthcare including clinical practice, medical education, governance, management and academia. He is therefore interested in reflexively situating academic bioethics as an aspect of social, cultural and political engagement with the moral and ethical questions of medicine, the biosciences and research.

Catherine Kendig is an assistant professor in the Department of Philosophy at Michigan State University. She earned her PhD at the University of Exeter/ESRC Centre for Genomics in Society and her MSc in Philosophy and History of Science at King's College London. Kendig specialises in philosophy of biology, philosophy and history of science, and metaphysics. Her main research interests are in philosophy of scientific classification, natural kinds, philosophy of science in practice, synthetic biology and philosophy of race. Her recent research in the metaphysics of synthetic biology has been funded by the National Science Foundation Division of Molecular and Cellular Biosciences (NSF-MCB). She is editor of the recent collection of interdisciplinary essays *Natural Kinds and Classification in Scientific Practice* (2016, Routledge).

Mike Michael is a sociologist of science and technology, and a professor in the Department of Sociology, Philosophy and Anthropology at the University of Exeter. His research interests have touched on the relation of everyday life to technoscience, the role of culture in biomedicine, and the interplay of design and social scientific perspectives. Recent major publications include *Actor-Network Theory: Trials, Trails and Translations* (Sage, 2017). He is currently writing books on science and technology studies and design (with Alex Wilkie) and on speculative research methodology.

Alexis Paton is a research associate with the SAPPHIRE (Social Science Applied to Healthcare Improvement Research) group at the University of Leicester. Her research interests focus on examining decision making in the medical community through the sociological and bioethical lens, as well as promoting the use of sociology and the social sciences in bioethics research.

Brian Rappert is a professor of Science, Technology and Public Affairs at the University of Exeter. His long-term interest has been the examination of the strategic management of information, particularly in relation to armed conflict. His books include *Controlling the Weapons of War: Politics, Persuasion, and the Prohibition of Inhumanity*; *Biotechnology, Security and the Search for Limits*; and *Education and Ethics in the Life Science*. More recently he has been interested in the social, ethical and political issues associated with researching and writing about secrets, as in his books *Experimental Secrets* (2009), *How to Look Good in a War* (2012) and *Dis-eases of Secrecy* (2017).

Hauke Riesch is a lecturer in sociology at Brunel University London. His research interests span the sociology and philosophy of science and science and risk communication. His current work includes examining the use of humour and comedy in science communication practice.

Nancy Tamimi is a dentist whose career moved to public health and medical sociology. She received her MSc in the science of Dental Public Health from Queen Mary University of London. Her PhD in Medical Sociology was awarded from Brunel University London. Currently, she is a teaching fellow in the Global Health and Social Science Department at King's College London. Her research interests include smoking, health inequalities; tobacco harm reduction; global health; experiences of health and illness and bioethics. Nancy is an associate fellow for the Higher Education Academy (HEA) in the UK, a member of the British Sociological Association (BSA) and the Society for the Study of Addiction (SSA).

Steven Wainwright is a qualitative (medical) sociologist with an unusual background in the social, earth and biomedical sciences and in the world outside academia. He worked in intensive care (Charing Cross Hospital, London) and taught intensive care nursing (Royal Free Hospital, London) before joining King's College London in 1995, where he held posts as lecturer, research fellow, senior lecturer and professor. He worked outside academia for around 15 years between his BSc and his first post as a university lecturer. He joined Brunel University London in 2011 as professor of Sociology of Science, Health and Culture.

Clare Williams is a medical sociologist and honorary professor of Medical Sociology at Brunel University London whose research has focused for the past 17 years on the intersection of sociology and ethics. Her interdisciplinary work has focused primarily on the ethical dilemmas of innovative medical technologies, particularly in the fields of embryonic stem cell research, in vitro fertilisation and preimplantation genetic diagnosis.

Introduction: Crossing the Divides

Hauke Riesch, Nathan Emmerich, and Steven Wainwright

1 Introduction

Since its inception, bioethics has been a multi-disciplinary area of inquiry. However, as Jonsen's (1998) history of the field shows, there have been a series of distinct periods where one or other of its contributing disciplines has come to the fore. If his account is to be believed theology, law and philosophy have all taken their turn at the forefront of the field. However, this pattern has not continued and (applied) philosophy or 'philosophical ethics' has been the predominant discourse within the field for some time. Furthermore, whilst they all make significant contributions to the field, the disciplines of history, sociology, and anthropology have not had their day. The most obvious distinction between these disciplines and those that have previously taken the lead in the field is that the disciplines of history, sociology, and anthropology all have significant empirical components.[1] Perhaps, then, it is the

---

[1] Subsequently, this introduction focuses on sociology rather than history or anthropology. In part this is because the clash between sociological and philosophical approaches is the one that is most obvious in the field of bioethics. It is also because the essays in this collection are primarily sociological in focus. However, it should also be read as a short hand for broader perspectives, including those of anthropology and history, in much the same way that Social Studies of Science (SSS) might be equated with Science and Technology Studies (STS). Whilst the latter more obviously includes anthropological and historical work, it would misguided to think that the former excludes them;

H. Riesch (✉) · S. Wainwright
Department of Social and Political Sciences, Brunel University London, London, UK
e-mail: Hauke.riesch@brunel.ac.uk; Steven.wainwright@brunel.ac.uk

N. Emmerich
ANU Medical School, Australian National University, Canberra, Australia

The Institute of Ethics, Dublin City University, Dublin, Ireland

School of History, Anthropology, Politics and Philosophy,
Queen's University Belfast, Belfast, UK
e-mail: nathan.emmerich@anu.edu.au

© Springer International Publishing AG, part of Springer Nature 2018
H. Riesch et al. (eds.), *Philosophies and Sociologies of Bioethics*,
https://doi.org/10.1007/978-3-319-92738-1_1

empirical disciplines as a whole that are taking their turn at the forefront of bioethics. Certainly, shortly after the turn of the millennium, Borry et al. (2005) hailed the birth of empirical turn, a phrase presumably chosen so as to echo the title of Jonsen's (1998) book, The Birth of Bioethics. However, even as such empirical scholarship – and the very notion of 'empirical bioethics' and 'empirical ethics' (Ives et al. 2016) – has proliferated, there remains a certain degree of scepticism regarding its value and, in particular, the relationship it has to philosophical bioethics, as well as the validity of its meta-ethical commitments.

In this context it would, of course, be myopic not to acknowledge the fact that (applied) philosophy continues to dominate both the intellectual discourse of bioethics and the field as a whole. Whilst one might think that this is not necessarily an inappropriate way to structure bioethics (Emmerich 2015a) it is nevertheless the case that much remains open for debate. However, both 'empirical ethics' and the empirical turn continues to be attract a certain amount of scepticism – not least of which is scepticism regarding the advent of any such thing as a 'turn' (Hurst 2010), rather than a greater degree of attention being paid to empirical scholarship that has long been part of bioethics. Thus, what seems most vital at this point in time concerns the relationship between discourses that present a philosophically normative approach and those that pursue more critical perspectives. This is to say that the concern is with the relationship between applied ethics, theological perspectives, jurisprudence and the 'black letter' leanings of medical law,[2] on the one hand, and sociological, anthropology and historical forms of analysis, on the other.

However, as this suggests, this should not be taken as a concern for the relationship between 'theoretical' or 'analytic' disciplines and those that are merely 'empirical' or 'handmaiden' to normative ethical analysis (Haimes 2002). Rather, it is with two different analytic modes. One the one hand we have those that are normative (or, in some minimal sense, 'action guiding' or structuring) and those that are critical and require a greater degree of discursive engagement if their ethically normative implications are to be realised. As such, the concern at hand is with the nature and appropriate meaning attached to interdisciplinarity in bioethics, or so it seems to us.

Certainly, a number of authors have previously devoted their attention to such concerns (De Wachter 1982; Ives 2014; Haimes 2002) and the essays contained in this volume continue this debate. However, arguably, they go further and seek to more fully embrace bioethics as an interdisciplinary form of inquiry. As such many of the essays draw on the conceptual resources and substantive insights of Science and Technology Studies, a field that has pioneered the kind of basic interdisciplinarity that bioethics still struggles to attain. Bioethics is not, of course, alone in struggling to come to terms with interdisciplinarity. In academia, work that crosses the broad boundaries of philosophical and sociological studies has often encountered problems, particularly in the case of philosophical and sociological studies of sci-

---

both of are interdisciplinary fields regardless of the nomenclature used to refer to them. The same can be said of interdisciplinary bioethics as well as the travails of sociology both in and of bioethics.

[2] On the relatively positivist inclinations or presumptions underpinning the relationship between medical law and philosophical bioethics, see: Harrington (2017).

ence (Riesch 2014). Agendas in philosophical bioethics and social scientific bioethics offer an example of the kind of conflict that can arise, particularly when one takes the fields of medical sociology and the anthropology of (bio)medicine into account. Clearly, each of these disciplines has much to offer and each has well developed and explicitly debated arguments about their relationship with broader discourses. Nevertheless, achieving an integrated and cohesive approach to interdisciplinary bioethics seems beyond the immediate grasp of this scholarly community, which is marked by disciplinary relationships that are somewhat strained.

These strained disciplinary relationships have the potential to hinder academic scholarship, which could otherwise benefit from a more interdisciplinary approach, one that captures the essence of both philosophical and sociological insights. Indeed, the feeling that bioethical debate has suffered from insufficient integration between the traditions seems remarkably persistent (Hedgecoe 2004; Paton 2017). Unfortunately this has led to misunderstandings, re-invented wheels and missed opportunities for collaboration. It also follows the pattern of debates about the inherent virtues of interdisciplinary research more generally. Broader arguments suggest that interdisciplinary research can help by encouraging new ways of looking at established problems through the insights that outsiders can bring, and bringing together researchers with complementary sets of skills (Nissani 1997). It enables the pooling of resources and it also has the potential to bring individual researchers more visibility to those who are able to address two or more academic communities. There is, then, a clear motivation for breaking away from the tensions of interdisciplinarity, and to foster dialogue between disciplines, so as to draw out areas of complementarities.

We do not, of course, wish to give the impression that inter-disciplinarity, or its virtues, are unproblematic (cf. Frickel et al. 2016). As Jacobs and Frickel (2009) argue, many of the statements on the supposed benefits of interdisciplinarity rest on little empirical evidence. Nor would we argue that all research should accommodate such an approach; rather we believe that it is the hostility that characterises some of the relationships between normative bioethics, sociological studies of bioethics – as well as Science and Technology Studies (STS) and philosophy of science more broadly – can be detrimental to our efforts of conceptualising and answering the often complex set of bioethical problems these disciplines set out to address.

This collection is therefore aimed at bringing together social scientists and philosophers in an effort to broaden the conversation across the disciplinary divide. Social science here comprises both sociologists of bioethics as well as STS scholars more broadly, whilst philosophy comprises both normative ethicists as well as philosophers of science more broadly. Here, there are clearly a series of complex and interwoven interdisciplinary relationships at work that defy superficial division of the debate into two distinct camps, and which makes clear that developing a strategy for communication between and across disciplinary boundaries is difficult. An important first step to developing a better relationship is the creation of mixed fora – such as this collection of essays and the conference that preceded it – in which scholars from the various disciplines present their work and discuss it with each other. We seek, therefore, to create a dialogue that embraces disciplinary similarities through the joint presentation of our work, alongside respecting (rather than rejecting) the very real disciplinary differences. Our aim is to promote a dialogue that understands and accepts both the benefits and the drawbacks of working across disciplines, and one that is acutely aware of when it may be productive to work together, and when it might not be.

This introduction will be our attempt at outlining where the disciplines of bioethics sit with regard to each other, what the main issues might be when sociologists, philosophers and STS scholars interested in bioethics encounter one another, as well as the challenges that we might face in trying to bring them together. While we are generally keen on interdisciplinary interactions (otherwise we wouldn't have put together this volume) and whilst believe that more is to be gained by working together than apart, interdisciplinarity also has its challenges; challenges that we feel are too often glossed over, particularly in debates about 'empirical ethics.' First, interdisciplinarity often lacks clear definition, as do the wished for positive consequences. Critical research on interdisciplinarity is still relatively rare and, rather than being based on strong empirical evidence, drives to enhance interdisciplinarity are often motivated by 'institutional,' 'managerial' or 'corporate' priorities rather than those of academics or the research being pursued. This is something that contributes to often ill-defined aims behind generalised attempts to promote interdisciplinarity. Thus, after a brief introduction to the debate regarding philosophical and sociological approaches to bioethics, we follow up with a more in-depth look at interdisciplinarity, its potential side-effects and, subsequently, how these might be understood in relation to bioethics.

2   Philosophical and Sociological Approaches in Bioethics

Since the mid 1980s, if not before, bioethics has come to be dominated by philosophical approaches and, in particular, by applied analytic approaches to ethics. For the most part, these involve the use of abstract, or 'arm-chair', reasoning to debate and/ or resolve questions of 'how moral agents should act' in the course of medical practice and when conducting research in the arena of the biosciences and biomedicine. As Lopez suggests, philosophical bioethicists tend to favour "a formalistic, procedural, disembodied and universalistic way of identifying and resolving bioethical dilemmas" (2004: 878). The assumption is that the insights generated by this approach can directly impact or (re)structure what is done in the clinic or during the course of conducting research. This ethical practice is achieved via the 'top-down' application of a set of principles and analytic ethical insights to specific clinical and research contexts (Hedgecoe 2004).

Beauchamp and Childress' (four) Principles of Biomedical Ethics (2009) has become the most well-known, and often criticized, example of this approach to ethics. Significantly influenced by Beauchamp's work on the National Commission for the Protection of Human Subjects of Biomedical and Behavioral Research in the 1970s, the four principles have become a short hand or conceptual vocabulary for discussion of ethics in medical practice, particular by professionals. Nevertheless, a majority of bioethicists, particularly philosophical bioethicists, are largely critical of 'principlism' or, as the particular specification of Beauchamp and Childress' four principles have been termed, the Georgetown mantra (Walters 2003: 225). Whilst he argues that the four principles have their virtues, particular in the pragmatic context of actual clinical practice, Huxtable identifies four particular criticisms within the literature. These are that they are imperialist; inapplicable; inconsistent; and inadequate. Finally, whilst some have promoted respect for autonomy as the first of four equal principles (Gillon 2003) others see the dominance that same principle as problematic and identify a theoretical lack in the methods for specifying and balancing the principles in concrete cases (Holm 1995). Nevertheless, the concern levelled at the four principles is not simply focused the creation, articulation and use of substantive principles, but with a range of other presuppositions and assumptions regarding the proper way to approach the ethics and the ethical analysis of practices such as medicine, healthcare and bioscientific research.

Proponents of philosophical bioethics and of applied ethics more generally, also tend to be sceptical of sociological input into the discipline. The reason for this can be traced to a fundamental commitment to the principles of analytic moral philosophy on the part of applied ethics (Glock 2011). In particular we might note the influence of the conceptual divide between "what is the case", which they view as the domain of sociology, and "what ought to be the case", which they view as the domain of philosophy (Humphreys 2008). For these scholars, Hume's "naturalistic fallacy" is being committed whenever a sociological researcher tries to do bioethics by inferring an "ought" from an "is" statement. Indeed, Humphreys points out that sociologists of bioethics "seem determined to violate this 'naturalistic fallacy' [...] by showing how things are – the 'is' – and then progressing to conclude that perhaps things are not as they ought to be" (Humphreys 2008: 49). One might counter, however, that philosophical ethicists tend to reason about how things ought to be, without pausing to consider whether, and how, what is the case can be reconfigured in the manner prescribed.

Of course, social scientists working in the area of bioethics have rejected such criticisms, not least because it fails to grasp the critical dimension of sociological analysis. Furthermore, those who pursue such sociological forms of analysis are critical of the influence of moral and analytical philosophy in bioethics. Drawing on empirical research, social scientists have argued that there is an incommensurability between the abstract, formal principles developed by bioethicists, and the world as it is actually experienced and practiced by human agents (Hoffmaster 1992; Kleinman 1999; Corrigan 2003; Hedgecoe 2004; Bosk 2009). The approach drawn on by philosophers, these scholars say, cannot account for the complex intertwining of social and cultural features that inevitably structure how moral dilemmas are perceived and negotiated by human agents in specific cultural environments, such as that of the biosciences and biomedicine. According to Hedgecoe's characterization of the argument, there is, a "significant difference between ethics as presented in bioethics, and the way in which ethical reasoning takes place in the clinic, as shown by an increasing number of sociological and anthropological studies" (Hedgecoe 2004: 121). Consequently, the formal principles of bioethics are often of little use in biomedical contexts or, at least, they would be of little use were they to be taken up in the manner prescribed by applied philosophy. Indeed, they can have a detrimental effect and produce a 'myopic' reliance on principlism of the type that has had negatively affected the practice of healthcare (Fox and Swazey 1984) and, one might add, social research (Schrag 2010; Van Den Hoonaard 2011).

Several social science commentators have argued that bioethics can be saved from irrelevancy if it were to more fully embrace methods from the social sciences (Hedgecoe 2004; DeVries and Conrad 1998). As it directs our attention towards the activities that constitute everyday life, ethnography is often singled out as being of particularly use or insight (Hoffmaster 1992; Kleinman 1999; Bosk 2001). Ethnographic research enables an exploration of the various ways in which cultural scripts, narrative techniques, institutional routines and relations of power can shape how agents perceive and negotiate ethical challenges in specific biomedical contexts (Kleinman 1999). Proponents argue that ethnographically-informed work has so far made several valuable insights into the nature of ethical work and that these insights further demonstrate the misguided nature of a top-down approach to ethics like applied principlism (Hoffmaster 1992: 1425).

In spite of this there still exists resistance towards closer collaboration, within both philosophical and sociological bioethics. For example, Goldenberg argues that "[i]ncreased integration of social and life scientists into the field [...] represents a loss of confidence in the typical normative and analytic methods of bioethics" (2005: 1). By focusing on the empirical "bottom-line", she argues that the 'normative mandate' of bioethics is under threat. Whilst Goldenberg's conclusion is motivated by concerns about the seemingly neutral technical measures offered by some forms of 'empirical ethics,' there are other factors that keep philosophical and sociological bioethics at arm's length. In his analysis of the interaction between the two communities Pickersgill suggests that "influential individuals do not always take kindly to having the 'black box' of their work unpacked and its contents inspected" (2013: 34). Pickersgill likens the tensions generated by these conflicts to the "science wars" in the 1990s. Similarly, in a paper bearing the title 'ethics wars,' Hoeyer notes that philosophical "bioethicists do not feel invited to a dialogue" (2006: 204) by sociologists of bioethics.

A number of bioethical scholars have paid close attention to social scientific critiques of the field, with at least some trying to address these concerns by conducting 'empirical' bioethical research. This field of research, which uses social science methods to explore ethical issues and then draws on philosophical concepts to discuss them, has gained a significant degree of support in the field. However, such integrative methods are still rejected by many, both from the disciplines of philosophy as well as from those of the social sciences. Some social science scholars argue that 'empirical ethics' misses the point: it is not that social science research needs to 'contribute' to moral and analytical approaches to bioethics via the use of empirical methods, but rather, that bioethics can be conducted via social science approaches in their own theoretical right (Paton 2017).[3]

---

[3]The recently resurgent fields of the sociology of morality and the anthropology of ethics offer further demonstration of this point. The viability of this work is independent of moral philosophy

Whilst, over the last decade or so, empirical bioethics has undergone significant development (Ives et al. 2017), it is still the case that it does not provide 'the answer' to the mutual irritations of philosophy and sociology. A number of philosophical and social science scholars have, however, started to explore the interrelatedness of their research. They do so from a starting point of mutual respect, acknowledging that both disciplines can make contributions to bioethics in their own right. Similarly, perspectives developed within STS are increasingly being used to analyse ethical debates, resulting in, for example, a number of articles that take up the notion of "boundary-work" (Wainwright et al. 2006; Frith et al. 2011; Hedgecoe 2001). Research from the Wellcome Trust funded LABTEC project has aimed to bring together scholars in normative and empirical bioethics as well as STS. It is, then, clear that there is no need to settle the nature of bioethics; lack of clarity is no barrier to the production of high quality disciplinary and interdisciplinary research. Nevertheless, there is value in examining 'bioethics' and doing so through various lenses, including notions of disciplinarity, interdisciplinarity and more sociological definitions. It is to this task that we now turn.

3 Interdisciplinarity as a Sociological Phenomenon

The virtues of being interdisciplinary have been rehearsed both often and forcefully. Nissani (1997), for example, gives a list of ten reasons to be cheerful about it. These include the idea that disciplinary "immigrants" can make new and unexpected contributions, that some errors can best be detected by people with two disciplinary backgrounds, or that some interesting topics might fall between the competences of individual disciplines. Mostly however, these virtues are not argued for in the scientific, social scientific or philosophical literature, but instead find themselves aired in policy documents, university administration rationales for restructuring exercises, research funders' calls for projects and similar places (Jacobs and Frickel 2009). Nissani's piece stands as a case in point. It is primarily a theoretical, rather than empirical, contribution. The arguments make sense and are convincing but, nevertheless, are not tested against any empirical research, other than some discussion of examples from the history of science.

The extent to which any of the envisaged ideals for interdisciplinarity is practically realisable is, despite the general enthusiasm surrounding it, very unclear, and barely any studies are conclusive. Evidence is hard to get by because the hopes and expectations invested in interdisciplinarity are often not very well defined or conceptualised. In a recent collection of papers on interdisciplinarity, Frickel et al. (2016) try to distil arguments about the virtue of interdiscipliarity into three main categories: "Interdisciplinary Knowledge is Better Knowledge"; "Disciplines

---

or applied ethics and whilst some of this work directly draws on virtue ethics, the theoretical dimension is fully anthropological, rather than philosophical (cf Laidlaw 2013).

Constrain Interdisciplinary Knowledge"; and "Interdisciplinary Interactions are Unconstrained by Hierarchies". However, each of these assertions can, in some way, be seen as problematic, or so they argue. This is either due to lack of clarity about what, for example "better knowledge" might mean, or because there is a lack of empirical evidence that shows these assertions to be true.

The present volume is itself intended to be interdisciplinary in scope. This introduction is not, therefore, intend to provide an argument against interdisciplinarity as such. However, it is nevertheless important to investigate what exactly we are doing – and what we risk creating – when we aim for more interdisciplinary interactions, and where philosophers, sociologists and STS scholars think and write about bioethics through efforts like this volume. Two themes in particular are worth highlighting, which we will then aim to link back to the disciplinary relations between philosophical and sociological studies of bioethics; appropriately, we think, one is more of a philosophical point whilst the other is more sociological.

The first of these issues is that, if we want to benefit from interdisciplinarity, we need to think about the particular disciplines that are being bridged. Are there differences between the various disciplines that would make interdisciplinarity between them differently valuable? It is at least plausible that interdisciplinarity would solve different problems depending on whether two natural scientific disciplines are being asked to work together, or two social science disciplines, or two humanities disciplines – or any combination thereof (glossing over for the moment that even within this coarse but traditional partition of academic disciplines there is a great variety of ways of doing, knowing and thinking). All these different disciplinary groupings tend to have different philosophical foundations. This applies to their methods and aims, the standards of what constitutes knowledge and to the differently evaluated status of evidence, laws of nature, explanation, and so forth that builds up the disciplinary core. Philosophical arguments on the benefits of interdisciplinarity therefore need to be tailored to the specific disciplines under question, or so we would suggest. Related to this point is that we also need to consider what is being lost when disciplinarity is abandoned. With all the enthusiasm for interdisciplinarity it may also be well worth considering that disciplines serve useful intellectual and organisational purposes. There are socio-historical reasons for their development (Wellmon 2015). In the course of creating and developing interdisciplines we therefore need to be careful not to throw out the good as well as the bad things. Thus, one final question here would be, what would we potentially lose through interdisciplinarity?

The second of the themes we wish to highlight is that, as valuable as such discussions undoubtedly are, interdisciplinarity cannot be properly understood from an abstracted, philosophical point of view alone. We should recognise that disciplines are socio-cultural and, perhaps more importantly, *political* phenomena. They are populated by human beings and contain sub-groups, with socially contingent norms, hopes, dreams and enmities. As in any other sphere of human activity, academics have social identities, constructed around modes of thought, ontological and epistemic commitments, and ways of evaluating evidence. Power relations are central to the way in which such groups work. As Callard and Fitzgerald (2015: chap. 6) point out, interdisciplinary work provides an environment within which such relations should be seen as having added significance. Important amongst these are: group influence within the university or wider academic worlds, with some disciplines finding it easier to have their views heard when other disciplines are fighting for survival. Forcing interdisciplinarity between disciplines that historically view each other as rivals, or where one more powerful discipline suffers the involvement of the other discipline only reluctantly and at the behest of management – or funders – risks failure, no matter what the philosophically justified benefits of pooling resources might actually be.

The two points above, one philosophical and one sociological, are intertwined. Evaluating or promoting an interdisciplinary relation between two or more disciplines needs to take into account the forms of knowledge production they are engaged in (and what would potentially be lost through collaboration) as well as any historical enmities, power differentials and other social relations between the disciplines. Quite obviously, rather than flowing from grand, generalised, and, therefore, often vague promises about the virtue of interdisciplinarity, this would need to be tailored to the particular disciplinary and interdisciplinary formations under question. Somewhat disappointingly, especially for the purposes of the interdisciplinarity advocated in this volume, these grand promises tend to be made with case studies from the natural sciences in mind. However, collaborations that take place across this divide would seem to be more challenging than those that do not (Callard and Fitzgerald 2015). Nevertheless, interactions between and within the arts, humanities and social sciences often remain undiscussed. This itself may be a result of just one of those power differentials between disciplines. The epistemic success of the natural sciences often means that any interlocutors must first accommodate themselves to their ways of doing things. Doing so results in a good deal of self-analysis on the part of said interlocutors. The same cannot be said of those that understand themselves as meeting on the same or, at least, similar, terms.

In a welcome break from this trend, Callard and Fitzgerald (2015) have described their experience of interdisciplinary working. Respectively a geographer and a sociologist they collaborated on projects involving neuroscientists. While, on the whole, their contribution is about social scientists working with natural sciences, their briefer exploration of interdisciplinarity across social scientific disciplines is interesting. They observe that, despite them having met at an interdisciplinary workshop, they felt that them getting together was not what the workshop was about – it was about getting social scientists to work with neuroscientists. This, however, raises the interesting question regarding the potential differences between interdisciplinary relationships when the two disciplines in question are much closer related, conceptually and methodologically than is the case in instances where the science/arts-humanities-social sciences divide is being traversed. If there is a difference to how we should conceptualise interdisciplinarity between the great divides and the small divides, then these would need to be elaborated, as it is possible that solutions worked out for one won't apply to the other.

This being the case there is a question of how to understand philosophy – or applied philosophical ethics – and, therefore, the nature of interdisciplinarity in bioethics where it has a central and influential role. This is not the place to go into these matters in any degree of detail. However, it is worth noting that, arguably, analytic philosophy and applied ethics has more in common with 'the sciences' that it does with 'the social sciences' or, indeed, other humanities disciplines (Glock 2008: 160–63).[4] For the most part, the methodological stance adopted by applied ethics is one of individualism. Whilst biomedical knowledge tends to be given statistical articulation and, therefore, is given in relation to certain populations the nature of clinical medical practice is such that healthcare professionals are in the business of caring for and treating individuals. Thus, in analysing the ethics of medical practice it is easy to focus on particular cases, instances in which a doctor (or healthcare team) encounters an individual patient, and to do so in a manner that is not dissimilar to the clinical endeavour, something that involves framing the individual case in terms of scientific (or ethical) generalisations. Furthermore, a good deal of social scientific research rejects the distinction between facts and values. This is something that is maintained in both the natural sciences and applied ethics. Indeed, in the case applied ethics, the aforementioned distinction between *is* and *ought* occupies a central position in the discipline's analytic perspective.

As a result, one might think that the kind of interdisciplinarity one might find, or seek to develop, in bioethics not only crosses small divides between various social scientific and humanities disciplines, but that there is also a greater divide between these disciplinary (and interdisciplinary) endeavours and an applied ethics, a mode of thought that is conducted in the light of its genealogical antecedent, analytic philosophy. We must, at this point, put aside our discussion of interdisciplinarity and, building on the comments offered in the previous section, (re)turn to a more detailed examination of bioethics itself. In the following section we seek to outline how we feel interdisciplinarity plays out within bioethics.

4 Understanding Bioethics: Towards Interdisciplinarity

As suggested above, the forms of scholarship one finds within bioethics are both disciplinary and interdisciplinary in nature.[5] It is not alone in having this character; many other forms of enquiry that emerged and developed in the latter part of the twentieth century are similarly structured. As previously noted, this includes STS and gender studies, but further examples might also include Socio-Legal Studies (SLS), media studies, criminology, as well as the fields of cognitive science and, more recently, neuroscience (Callard and Fitzgerald 2015). The question we would like to pose is how best to understand these intellectual fields, and how they might

---

[4] The fact that there is an affinity between the disciplinary ethos of analytic philosophy and the natural sciences is not refuted by Glock's conclusion; that the 'scientific spirit' of analytic philosophy cannot provide a basis with which to define the field or the discipline (2008: 163).

[5] We leave the nature of transdisciplinarity, and any differences between this and interdisciplinarity, particularly in bioethics, to one side.

differ from (mono) disciplinary forms of enquiry, if indeed they do. In order to do so, we first consider the nature of a discipline – or disciplinarity – itself.

At first blush it would seem that the correct way to define a discipline is through a combination of the topic it takes as its focus (or, more philosophically, its ontology) and the epistemic and methodological approach it takes to examining or analysing that focus. Such a picture could be taken as justified by the presumption of realism and the notion that the disciplines (ought to) reflect the structure of reality. As such the development of disciplinarity would seem to reflect a kind of intellectual progress, one in which human kind was developing a better grasp of what one might call the natural order. However, within such a views, there is a distinct potential for a troubling degree of reductionist – and mechanistic – thought; it seems predicated on the notion that the hierarchical structure of the disciplines – in ascending order: physics, chemistry, biology, psychology, sociology – not only reflects the structure of reality, but that higher level phenomena, such as sociology, can be reduced to a lower or more basic level, such as psychology and, in turn, biology. It would also seem committed to certain other notions that have been subject to increasing pressure, such as the distinction between pure and applied research, science and engineering, or 'mode 1' and 'mode 2' forms of knowledge (Gibbons et al. 1994).

However, were this picture accurate then, outwith any minor adjustments to disciplinary boundaries, it would be difficult to see how or why any fully interdisciplinary forms of enquiry would subsequently arise. If the hierarchical structure of existing disciplines and the knowledge they produce 'carves nature at its joints' then why would any form of extra-disciplinary enquiry be required?

Furthermore, given that the same object (atoms, molecules, organisms, human beings, societies, planets, solar systems and so forth) can usually be examined from a number of different perspectives (physics, chemistry, biology, various social sciences, including sociology, criminology and psychology, and various humanities, such as anthropology and history) it would seem that the epistemic and methodological features of a discipline contribute to the definition of a discipline alongside any substantive or ontological focal point(s). Indeed, from one point, they could be could be taken as constituting the *discipline* per se. Whereas a disciplines' ontology merely constitutes the focus of its attention, methodological and epistemological issues define the correct way in which a particular enquiry should be pursued and, therefore, they represent the very *discipline* that constitutes the *practice*.

Nevertheless it is, we think, better to conceptualise a discipline as an interrelation of ontological, epistemic and methodological features. Whilst disciplines may, on the face of it, share objects of concern, the way in which they are conceptualised – and, so, the ontological presumptions they work with – varies in accordance with their epistemic and methodological requirements. From a theoretical point of view, the ontology of a discipline is not an external or independent *given* but, we might say, an *assemblage* which emerges from the disciplinary perspective as a whole. Furthermore, defining ones subject matters has implications on the way in which it should be approached and understood. The epistemological, methodological and ontological features of a discipline interrelate and they must, therefore, be taken together. In this light, a historically informed account might adopt the view that, following the enlightenment, and over the course of the eighteenth and nineteenth centuries, the need to 'breakdown' intellectual inquiry into specific areas, and discrete specialism, became increasingly apparent (Wellmon 2015). Thus, at least initially, various natural sciences gradually emerged from what was previously defined as natural philosophy and, somewhat later, psychology and the social sciences developed from what was 'the moral sciences.'

Thus, rather than reflecting natural kinds, a more defensible view of disciplines might be that they are, simply, socio-historical realities; intellectual subcultures that have arisen around particular forms of enquiry (epistemic and methodological practices) and the substantive focus of those forms of enquiry take (the metaphysical as well as ontological presumptions articulated within those practices). Such a picture would perhaps explain why it is that both the theory of relativity and quantum mechanics continue to be understood as belonging to physics, despite the radical differences in focus and methods, as well as their apparent incompatibility. Similarly, it would account for the fact that American anthropology includes biological, linguistic, cultural and archaeological approaches whereas its European counterpart tends to restrict its concerns to cultural or social anthropology alone. Such a view would also make room for the development of new forms of inquiry that adopt increasingly complex ontological and metaphysical perspectives, not least because such perspectives are understood to be entangled with, rather than prior to, the epistemic and methodological commitments of a particular discipline.

If, as would seem to be the case (Wellmon 2015), that the advent of disciplinarity was implicitly informed by the metaphysics of enlightenment thinking, then it would seem that there is something decidedly post-enlightenment about the metaphysics of interdisciplinarity.[6] Perhaps the clearest example of this can be found in the notion of the assemblage (Legg 2011) or in the move from structuralism to post-structuralism in the social sciences and in social studies more generally.[7] For some forms of interdisciplinarity the transition has been relatively unproblematic. This includes forms of interdisciplinarity in which there is little need to directly attend to the transition from enlightenment to post-enlightenment thinking; here one might think of biochemistry, or medical physics, as well as other arenas where two or more of the natural sciences are combined. It also includes forms of interdisciplinarity that actively embrace post-enlightenment thought, with STS being the most obvious example. Where difficulties arise is when an interdisciplinary field entails combining practices within which both enlightenment and post-enlightenment thinking can

---

[6] We use the terms 'enlightenment' and 'post-enlightenment' in preference to the more common 'modern' and 'post-modern.' This is largely due to a widespread misunderstanding of these latter terms as well as the reflex reaction they often prompt amongst those who labour under such misapprehensions.

[7] This is not, of course, to suggest that all those working within these fields have made such a transition. Rather it is to point out that the advent of post-structuralism opens up space for interdisciplinary work that was not previously possible. There is, we would suggest, some degree of correspondence between disciplinarity and structuralism, on the one hand, and interdisciplinarity and post-structuralism on the other.

be found. Given our comments on analytic philosophy and applied ethics, as well as on the increasing influence of sociology within the field, this is, we would suggest, the position bioethics now finds itself; something that has, arguably, been a long time coming.

As the history of bioethics shows, the field has always included a range of disciplines or forms of enquiry. Furthermore, how to define (and, therefore, discipline) the field (and its practitioners) has also been the subject of long discussions. However, even if the explicit view seems to be that bioethics was and could continue to be a 'big tent discipline' (de Vries et al. 2006: 676), the implicit presumption nevertheless seemed to be that, eventually, the field would become unified around a set of (inter)disciplinary norms, and thereby a range of settled, integrated and inter-related set of academic practices would emerge. Whilst it is clear that some working in the field continue to aim at the latter two goals (Ives et al. 2016), the best that can be said is that the academic practices that define the field have become settled, at least to a degree. However, the fact that we have been unable to properly interrelate and fully integrate such practices has contributed to the fact that the field has been unable to produce a settled academic identity. Thus, despite ongoing efforts to define bioethics, the way in which it ought to be pursued and the nature of various related endeavours (such as the clinical ethicist), conflict over the term, and to whom it might belong, remains an issue.

Somewhat ironically, those who refuse to embrace the term bioethicist best exemplify this point. Whilst some who work within the fields of sociology, anthropology and history presently embrace the term 'bioethicist,' they tend to do so as part of a strategy to reclaim bioethics from applied ethics. The majority of these individuals understand themselves and their work as primarily falling within, respectively, medical sociology, anthropology of medicine and the history of science, technology and medicine, but nevertheless see themselves and their scholarly work as contributing to bioethics. What is less commonly recognised – or, perhaps, acknowledged – is that many current applied ethicists – individuals who could be legitimately thought of as bioethicists *per se* – are either ambivalent about being seen as bioethicists, or they explicitly reject the label bioethicist, preferring to see themselves as philosophers.[8] For all its success – and there is no denying that bioethics has been successful, both in terms of becoming an institutionalised feature of the academy, attracting significant amounts of funding and undergoing significant intellectual development – bioethics can often seem to be a field that is divided against itself.[9]

---

[8] The most common claim is that they are philosophers who work within some combination of moral philosophy, metaethics, and applied ethics. However, we have also encountered those who consider themselves to be political philosophers. Our point here is not to deny these claims, merely to point out that they are taken up in preference to and, therefore, in rejection to being thought of as a bioethicist. Indeed, the claim that they are philosophers, rather than applied ethicists, is itself significant.

[9] In the US there is a further division between those who pursue bioethics as an academic endeavor and those who pursue it in a professional capacity, working within hospitals and offering clinical ethics consultations.

Of course, the dynamics of such claims are informed by more than just concerns that arise from the field of bioethics itself.[10] For example, in philosophy there is a line of thought that 'applied ethics' does not belong to the discipline and that even it if does, that it is a low status sub-field. As such there is a tendency to perceive those that work within applied (bio)ethics as not addressing – and therefore being unable to address – the really important and difficult philosophical questions. In this context, the attempt to maintain ones claim to philosophy, or to being a philosopher, can entail rejecting a claim to being a bioethicist. Similar, although less vitupera-tive, thoughts can be applied to sociologists, anthropologists and historians who contribute to bioethics; being seen as a bioethicists may come at the cost of no lon-ger being seen as a sociologist, an anthropologist or a historian.[11]

It would seem, then, that whilst appealing to the nexus of ontology, epistemology and methodology offers some guarantee that any enquiries conducted within a par-ticular discipline are appropriately structured, interdisciplinarity creates problems for the putative unity such a definition demands. One alternative is to define the inter/discipline of bioethics *sociologically* and through reference to its existence as an *intellectual field*. However, as the above comments suggest, we cannot proceed to define bioethics on the basis of who is thought to be a bioethicist and who is not. After all, whose view should we take on the matter? At the very least, a sociological investigation will need to take the published literature as providing some kind of initial picture.

However, even this might be a misguided way to go about things if we consider whether or not 'bioethics' is something that can be done in non-academic settings. First, it would not be unusual; a range of other academic disciplines are pursued in non-academic, but nevertheless, professional contexts. Consider, for example, architecture, engineering, medicine, and the (bio)scientific research currently being conducted by private corporations. Second, it would seem to be self evidently true that bioethics is pursued in non-academic domains; after all what else are those working within the USA's various President's Commissions on Bioethics or in the

---

[10] That said, the particulars of the field of bioethics, and the way it lays claim to individuals, are an important aspect of this dynamic. By way of comparison, consider the field of business ethics. Like bioethics, scholars from a range of disciplinary backgrounds contribute to the literature on the subject. However few, if any, consider business ethics to be a particular inter/discipline or even a specific (intellectual, which is to say, social) field, as opposed to a subject or topic. Indeed, the most significant book on the topic in recent years is a work of sociological history (Abend 2014). Thus, debate about the 'correct' way to approach the subject, or how critical sociological investiga-tions should be related to the normative work of applied ethics, is far less pronounced than it is in bioethical literature.

[11] A major factor in the production of this dynamic has been the fact that many individuals working in these areas have found employment within medical schools, or centers for biomedical ethics attached to faculties of medicine and the biological sciences, rather than in departments of philoso-phy, history, sociology and anthropology. In this context, appealing to the norms of ones field is part of justifying ones work and its value, particularly in the face of assessment exercises – such as the UK's Research Excellence Framework (REF) – and in applications for (internal) promotion. Obviously the narratives and, in particular, the metrics, associated with such endeavours are a major component of this.

UK's Nuffield Bioethics doing if they are not doing bioethics? Similarly, whilst it may be true that Clinical Ethics Consultants do things other than bioethics, such as mediating conflicts between patient's families and healthcare professionals, it would seem problematic to deny that what they are doing counts as bioethics. Indeed, given the importance of professional medical ethics to the field of bioethics it would also seem to be the case that what healthcare professionals do has, at the very least, a bioethical component.[12]

Seen in this light, it is not merely the case that bioethics has a practical *orientation*; it also has a practical *component*. As such, our initial presumption that bioethics should be defined as an academic inter/discipline, would itself seem to be misguided. This brings to the fore another bioethical fault line. Human beings are moral agents. However, few human beings study moral philosophy or applied ethics. Nevertheless, the *modus operandi* of these intellectual disciplines is the proper investigation of morality and ethics; the goal being to uncover the truth of the matter. This includes determining matters of right and wrong, good and bad, and – vitally – the correct way to determine such things.

The implication would seem to be that 'the correct way to determine these things' is through the pursuit of moral philosophy and applied ethics. It would seem, then, that when it comes to distinguishing right from wrong, and good from bad, the vast majority of moral agents are at a distinct disadvantage, at least as compared to the applied ethicist and the moral philosopher.[13] Concern for this point can be found in debates about (bio)ethical expertise. The matter can nevertheless be placed to one side if we consider that what it is to be a moral agent is to bear a form of moral responsibility that is inalienable. Thus, although laying claim to ethical expertise need not necessarily entail any such thing, deference to moral authority would seem to be deeply problematic from a moral point of view. Such thinking further reiterates a point we made above: given that the boundaries of ethics are not academic, neither are those of bioethics; they must, in principle, encompass a good deal of what ordinary moral agents do or are, in principle, capable of doing.

As such, some sociological contributions to the literature have discussed patients who encounter novel ethical issues as moral – which is to say *bioethical* – pioneers (Rapp 1988; Williams et al. 2005; Rose and Novas 2005: 460). Such individuals can be thought of as, in some way, 'doing bioethics.' If such thoughts are correct then the notion of bioethics as a discipline, or even a purely academic or intellectual field may be misguided. Perhaps, then, bioethics can be seen as a set, or sets, of ethical questions that arise and emerge from the life sciences –including climate science – and modern medicine and healthcare. This is not to return to an ontological definition

---

[12] We express this thought in these terms as we consider the idea that because there is a (bio)ethical dimension to (medical) practice does not make all those engaged in such practices a (bio)ethicist to be misguided. At minimum what is required is a certain degree of reflective and analytic engagement with the ethical dimension of practice if aspects of that practice are to be considered as rising to the level of bioethics.

[13] Of course, culturally speaking, this would not be an unusual state of affairs, even when it comes to moral matters. In the past theologians and priests could have been cast as those with privileged access to morality.

of a discipline precisely because such questions emerge from an unavoidable inter-action between science and society; bioethical questions cannot be framed in onto-logical terms alone, at the point at which they arise they already have an epistemological component. This is not to deny that the formal analysis of such questions is an important part of the field or that such analysis should cease to settle, define or otherwise fix its ontological presuppositions. It is only to suggest that the study of how such questions arise and are responded too, both within and without the academy, is an equally important component of academic bioethics. Furthermore, how bioethical questions and our academic and non-academic responses to them are taken forward, how they become incorporated into public policy, and, subsequently, how they shape future scientific and medical practices – and thereby give rise to and shape further bioethical questions – should also be considered of particular interest to both bioethicists as well as the field in general.

One might attribute not dissimilar thought to Brody when, in *The Future of Bioethics*, he says that "social scientists have … out reflected [the ethicists]" (Brody 2009: 36). In offering this remark Brody is acknowledges that empirical work does not simply offer up facts for normative evaluation. Rather, there is a critical, and therefore normative, purpose to the disciplines of sociology, anthropology and his-tory. Applied bioethicists have tended not to engage with the kinds of normativity that results from or is embedded within critical analysis. The way in which the 'hermeneutics of suspicion' is deployed mean that such perspectives are, one might say, crypto-normative. However, crypto-normativity need not be rejected (Kolodny 1996), it has purpose, value and can contribute to the development of bioethics. Whilst it may not establish any (applied) ethical certainties, it can challenge current practices and existing arrangements, as well as result in positive change. If bioethics is to become more fully interdisciplinary, particularly in the sense of seeking further cooperation and integration between disciplines, then we will need to engage with the question of cryptonormativity and with an academic approach to ethics that does not aspire to formal objectivity and abstract universalism. Rather, what is required is a bioethics that more fully acknowledges itself as starting from a particular social, cultural and political location; as being rooted within a particular standpoint that can itself be brought into question and considered from an ethical or evaluative point of view.

5   The Essays Contained in This Volume

This volume is intended to contribute to the debates and issues outlined above through the contributions of scholars situated in bioethics, sociology of bioethics, science and technology studies and philosophy. Whether all these descriptions ade-quately capture how we all individually feel about our place in the academy is of course, as the preceding discussion has attempted to show, difficult to say. Nevertheless, all the contributed chapters capture the spirit of crossing divides through either elaborating theoretical concepts developed in a broader discipline like STS and HPS and applying them to bioethics or bioethical case studies (Emmerich, Kendig, Rappert) or providing case studies of sociological research that are then applied to discussions in bioethics and risk (Riesch, Tamimi, del Savio, Paton and Wainwright et al.). Together they present an attempt at drawing these discussions together and showing how bioethical contributions that seek to cross the divides can look like.

In his chapter, *Emmerich* makes use of the perspective developed by Collins and Evans' as part of what they call the "third wave" of science studies. In particular he attends to the notion of expertise, and the concept of elective modernism. Seeking to reign in the excesses of second wave science studies, Collins and Evans have developed their ideas as a way to defend the idea that scientific knowledge has a privileged role to play in democratic politics. Emmerich considers how we might understand the role of bioethics in the context of the democratic governance of the biosciences, biotechnologies and biomedicine. As the field is directly concerned with values and evaluative judgements, the issues that arise in the use of expert testimony are more problematic when it comes to bioethics as opposed to science. Drawing on his previous analysis, Emmerich advances arguments he has previously advanced regarding bioethical expertise (Emmerich 2015b, 2016) considering how it should be appropriately exercised and made use of in democratic fora.

Kendig looks at how both metaphysical and epistemological questions, and not just ethical questions, arise within synthetic biology research. From the science-in-practice perspective as developed within integrated History and Philosophy of Science research (see Chang 2004), Kendig investigates the specific example of reengineering of metabolic pathways through systems biology and the attendant ethical problems that arise at the intersection of science, philosophy of science and sociology of science. She uses this example to sketch out an "extended agency ethics" that emphasises the intersubjective nature of her approach and where the "responsibility for the form of ethical evaluations is distributed across the system of agents within the network of practitioners".

The ethical issues that arise from biomedical citizen science reflects on the ethical issues are being examined by *Del Savio*, where citizen science refers to scientific research projects that involve the contribution of members of the public (see Riesch 2015). He reconstructs the ethical debates that underpinned the example project of uBiome and uses this case study as an example of how the use of Science and Technology Studies can enhance ethical analysis as it allows the researcher to identify a broader set of social goals and adds critical reflection on stakeholder views.

By investigating how and in which circumstances ethical issues are missing, ignored or unacknowledged, *Rappert* looks at what is not being talked about. This he argues, shows that we can ask pointed questions about our values and the limits of ethical reasoning. This, he argues, allows us to attend to the limits of both social and ethical analysis and thus find value in getting these disciplines to work together.

Paton uses her research into older women's experiences of making oncofertility decisions as a case study to develop an argument on how bioethics research can be conducted through sociological analysis. She argues that this type of research will enhance our understanding of the efficacy of existing bioethical theories which, in turn, will help better inform policy and clinical decision making. Thus her work shows how bioethics can benefit from sociological work and keep it current.

By introducing Geoffrey Rose's "prevention paradox" through a case study on newspaper framings of public health advice on drinking during pregnancy, *Riesch* argues that the different horns of the paradox are represented by different ways of framing public health advice. Thus, efforts of promoting public health can reasonably be ignored by individuals because individual benefits do not clearly align with public health benefits. This then shows how ethical considerations on a population level can take a different aspect when considered individually and therefore how ethical interventions need to be tailored to how they are understood by the supposed beneficiaries.

With her chapter on "vaping", *Tamimi* presents a qualitative study of users and health advisors views on electronic cigarettes as well as the regulatory development of this still fairly new technology in the UK. By linking the bioethical framework of principlism to the harm reduction principle in public health studies, she discusses the ethical issues raised by nicotine addiction and the use of electronic cigarettes to combat it.

Wainwright, Michael and Williams explore how risk is being enacted by liver disease stem cell therapy clinicians. Based on a series of qualitative interviews, they trace how ethics is being enacted by the clinicians as being caught the "rock of courage" and the "hard place of caution". They suggest that risk can be treated as a resource to understand the social world inhabited by medical and biomedical actors.

6 Conclusion

In a recent American Journal of Bioethics target article, Lee (2017) argues that the currently ascendant field of public health ethics can act as a bridge between contemporary bioethics and environmental ethics. Furthermore, whilst these two fields are presently distanced from one another, both fall within the scope of public health ethics, as a sub-field of bioethics. Those who coined the term bioethics envisioned a form of enquiry that was far broader than what currently take place under its rubric and, in both substantive and disciplinary terms, present day bioethics arguably neglects the bigger picture.[14] This is, as Lee (2017) suggests, most obvious insofar as discussions of climate change, for example, are not understood as being 'bioethical,' at least not by most of those who work in bioethics and environmental ethics.

The conceptual origins of the field are, then, such that they entailed enquiries of a more encompassing type than is currently the case and, indeed, it was broader than

---

[14] The term bioethics was independently coined at least three times, the first of which took place in Germany, in 1927, and was by Fritz Jahr (Sass 2008; Jahr 1927). Subsequently, both Andre Hellegers and Van Rensselaer Potter coined the term circa 1970, meaning that the birth of American bioethics was 'bi-located' (Jonsen 1998: 27; Reich 1994, 1995). In all three cases, the vision for the field of bioethics was broader than the way the field is understood today.

the particular expansion pursued in this volume as a whole, as well as suggested by the previous remarks we have made in this introduction. Whilst some of the chapters in this collection have attended to matters of public health, our collective purposes has been to more fully develop the relationship between philosophical and sociological forms of enquiry within the field of bioethics. Lee's intent is to show how public health ethics can bridge between contemporary bioethics and environmental ethics; how public health ethics can fulfil the visions for the field proposed by its originators. The argument she presents is, in essence, yet another call for a greater level of interdisciplinarity for a field that has consistently claimed to be interdisciplinary and even *inherently* interdisciplinary (Miller 2017: 14).

Our point is not that bioethics has failed to fulfil its ambition or that the claims it makes are in some way false or merely rhetorical. Rather it is that interdisciplinarity is not easy; there is always more that could be done, and further avenues that could be pursued. As we have tried to show, attempts to forge interdisciplinarity are bound up in disciplinary and inter-disciplinary politics, as much as epistemological, methodological and metaphysical disagreements. To address these issues we have sought to confront directly what issues are in play when bioethicists of whichever disciplinary persuasion engage in boundary work. These include the issues written about in the literature we have reviewed, such as worries over sociologists committing the naturalistic fallacy, or bioethicists failing to address real life concerns of the stakeholders through overly detached moralised reasoning. Attempts at "crossing the divides" like the present volume clearly won't be able to resolve these issues by themselves. However, we hope to have been able to show that there are social and political issues at play that certainly warrant being aware of.

Acknowledgements This collection arose from a one-day workshop held at Brunel University London in May 2013. Thanks are due to the Wellcome Trust funded LABTEC project and, in particular, to Professor Clare Williams for supporting that workshop, financially and morally, and Erika Mansnerus for helping with the organisation. Thanks also to Gabby Samuel and John Gardner for helping with an early draft of this introduction. Special acknowledgement, finally, to Amanda Rohloff who helped with the organisation of the early stages of the workshop but sadly passed away before it was held – this volume is dedicated to her memory.

References

Abend, G. 2014. *The Moral Background: An Inquiry into the History of Business Ethics*. Princeton: Princeton University Press.

Beauchamp, T.L., and J.F. Childress. 2009. *Principles of Biomedical Ethics*. 6th ed. Oxford: Oxford University Press.

Borry, P., P. Schotsmans, and K. Dierickx. 2005. The Birth of the Empirical Turn in Bioethics. *Bioethics* 19 (1): 49–71.

Bosk, C.L. 2001. Irony, Ethnography, and Informed Consent. In *Bioethics in Social Context*, ed. B.C. Hoffmaster, 199–220. Temple University Press: Philadelphia.

———. 2009. *What Would You Do?: Juggling Bioethics and Ethnography*. Chicago: Chicago University Press.

Brody, H. 2009. *The Future of Bioethics*. 1st ed. Oxford: Oxford University Press.

Callard, F., and D. Fitzgerald. 2015. *Rethinking Interdisciplinarity Across the Social Sciences and Neurosciences*. London: Palgrave Pivot.

Chang, H. 2004. *Inventing Temperature: Measurement and Scientific Progress*. Oxford: Oxford University Press.

Corrigan, O. 2003. Empty Ethics: The Problem with Informed Consent. *Sociology of Health and Illness* 25 (7): 768–792.

DeVries, R., and P. Conrad. 1998. Why Bioethics Needs Sociology. In *Bioethics and Society*, ed. R. DeVries and J. Subedi, 233–257. Upper Saddle River: Prentice Hall.

de Vries, R., L. Turner, K. Orfali, and C. Bosk. 2006. Social Science and Bioethics: The Way Forward. *Sociology of Health & Illness* 28 (6): 665–577.

De Wachter, M.A. 1982. Interdisciplinary Bioethics: But Where Do We Start? A Reflection of Epochè as Method. *The Journal of Medicine and Philosophy* 7 (3): 275–287.

Emmerich, N. 2015a. What Is Bioethics? *Medicine, Health Care and Philosophy* 18 (3): 437–441. https://doi.org/10.1007/s11019-015-9628-7.

———. 2015b. A Sociological Analysis of Ethical Expertise: The Case of Medical Ethics. *SAGE Open* 5 (2): 1–14. https://doi.org/10.1177/2158244015590445.

———. 2016. A Sociological Analysis of Ethical Expertise: The Case of Bioethics. *Cogent Social Sciences* 2 (1): 1143599. https://doi.org/10.1080/23311886.2016.1143599.

Fox, R.C., and J.P. Swazey. 1984. Medical Morality Is Not Bioethics: Medical Ethics in China and the United States. *Perspectives in Biological Medicine* 27 (3): 336–360.

Frickel, S., M. Albert, and B. Prainsack, eds. 2016. *Investigating Interdisciplinary Collaboration: Theory and Practice Across Disciplines*. New Bruswick: Rutgers University Press.

Frith, L., A. Jacoby, and M. Gabbay. 2011. Ethical Boundary-work in the Infertility Clinic. *Sociology of Health & Illness* 33 (4): 570–585. https://doi.org/10.1111/j.1467-9566.2010.01308.x.

Gibbons, M., Limoges, C., Nowotny, H., Schwartzman S., Scott P., Trow M. (1994). The New Production of Knowledge: The Dynamics of Science and Research in Contemporary Societies. London: Sage.

Gillon, R. 2003. Ethics Needs Principles—four Can Encompass the Rest—and Respect for Autonomy Should Be 'First among Equals. *Journal of Medical Ethics* 29 (5): 307–312.

Glock, H.J. 2008. *What is analytic philosophy?* Cambridge: Cambridge University Press.

Glock, H.J. 2011. Doing Good by Splitting Hairs? Analytic Philosophy and Applied Ethics. *Journal of Applied Philosophy* 28 (3): 225–240. https://doi.org/10.1111/j.1468-5930.2011.00538.x.

Goldenberg, M.J. 2005. Evidence-Based Ethics? On Evidence-Based Practice and the 'Empirical Turn' from Normative Bioethics. *BMC Medical Ethics* 6 (1): 11.

Haimes, E. 2002. What Can the Social Sciences Contribute to the Study of Ethics? Theoretical, Empirical and Substantive Considerations. *Bioethics* 16 (2): 89–113.

Harrington, J. 2017. *Towards a Rhetoric of medical law: Against ethics*. Abingdon: Routledge.

Hedgecoe, A.M. 2001. Ethical Boundary Work: Geneticization, Philosophy and the Social Sciences. *Medicine, Health Care and Philosophy* 4 (3): 305–309.

———. 2004. Critical Bioethics: Beyond the Social Science Critique of Applied Ethics. *Bioethics* 18 (2): 120–143.

Hoeyer, K. 2006. 'Ethics Wars': Reflections on the Antagonism Between Bioethicists and Social Science Observers of Biomedicine. *Human Studies* 29 (2): 203–227.

Hoffmaster, B. 1992. Can Ethnography Save the Life of Medical Ethics. *Social Science and Medicine* 35 (12): 1421–1431.

Holm, S. 1995. Not Just Autonomy—the Principles of American Biomedical Ethics. *Journal of Medical Ethics* 21 (6): 332–338.

Humphreys, S.J. 2008. The Sociology of Bioethics: The 'Is' and the 'Ought. *Research Ethics Review* 4 (2): 47–51.

Hurst, S. 2010. What 'empirical turn in bioethics'? *Bioethics* 24 (8): 439–444.

Ives, J. 2014. A method of reflexive balancing in a pragmatic. *Interdisciplinary and Reflexive Bioethics*. Bioethics 28 (6): 302–312.

Ives, J., M. Dunn, and A. Cribb, eds. 2016. *Empirical Bioethics: Theoretical and Practical Perspectives*. Cambridge: Cambridge University Press.

———. 2017. Theoretical Perspectives: An Introduction. In *Empirical Bioethics: Theoretical and Practical Perspectives*, ed. J. Ives, M. Dunn, and A. Cribb. Cambridge: Cambridge University Press.

Jacobs, J.A., and S. Frickel. 2009. Interdisciplinarity: A Critical Assessment. *Annual Review of Sociology* 35 (1): 43–65. https://doi.org/10.1146/annurev-soc-070308-115954.

Jahr, F. 1927. Bioethik: Eine Übersicht Der Ethik Und Der Beziehung Des Menschen Mit Tieren Und Pflanzen. *Kosmos* 24: 21–32.

Jonsen, A.R. 1998. *The Birth of Bioethics*. Oxford: Oxford University Press.

Kleinman, A. 1999. Moral Experience and Ethical Reflection: Can Ethnography Reconcile Them? A Quandary for "The New Bioethics". *Daedalus* 128 (4): 69–97.

Kolodny, N. 1996. The Ethics of Cryptonormativism: A Defense of Foucault's Evasions. *Philosophy & Social Criticism* 22 (5): 63–84.

Laidlaw, J. 2013. *The Subject of Virtue: An Anthropology of Ethics and Freedom*. Cambridge: Cambridge University Press.

Lee, L.M. 2017. A Bridge Back to the Future: Public Health Ethics, Bioethics, and Environmental Ethics. *The American Journal of Bioethics* 17 (9): 5–12. https://doi.org/10.1080/15265161.2017.1353164.

Legg, S. 2011. Assemblage/Apparatus: Using Deleuze and Foucault. *Area* 43 (2): 128–133. https://doi.org/10.1111/j.1475-4762.2011.01010.x.

Lopez, J. 2004. How Sociology Can Save Bioethics... Maybe. *Sociology of Health and Illness* 26 (7): 875–896.

Miller, F.G. 2017. Challenging the Conventional Wisdom: From Philosophy to Bioethics. *Perspectives in Biology and Medicine* 60 (1): 3–18.

Nissani, M. 1997. Ten Cheers for Interdisciplinarity: The Case for Interdisciplinary Knowledge and Research. *The Social Science Journal* 34 (2): 201–216.

Paton, A. 2017. No Longer 'Handmaiden': The Role of Social and Sociological Theory in Bioethics. *IJFAB: International Journal of Feminist Approaches to Bioethics* 10 (1): 30–49.

Pickersgill, M. 2013. From 'Implications' to 'Dimensions': Science, Medicine and Ethics in Society. *Health Care Analysis* 21 (1): 31–42. https://doi.org/10.1007/s10728-012-0219-y.

Rapp, R. 1988. Moral Pioneers. *Women & Health* 13 (1–2): 101–117. https://doi.org/10.1300/J013v13n01_09.

Reich, W.T. 1994. The Word "Bioethics": Its Birth and the Legacies of Those Who Shaped It. *Kennedy Institute of Ethics Journal* 4 (4): 319–335.

———. 1995. The Word Bioethics: The Struggle Over Its Earliest Meanings. *Kennedy Institute of Ethics Journal* 5 (1): 19–34.

Riesch, H. 2014. Philosophy, History and Sociology of Science: Interdisciplinary Relations and Complex Social Identities. *Studies in History and Philosophy of Science Part A* 48: 30–37. https://doi.org/10.1016/j.shpsa.2014.09.013.

———. 2015. Citizen Science. In *International Encyclopedia of the Social & Behavioral Sciences*, ed. J.D. Wright, vol. 3, 2nd ed., 631–636. Oxford: Elsevier.

Rose, N., and C. Novas. 2005. Biological Citizenship. In *Global Assemblages: Technology, Politics, and Ethics as Anthropological Problems*, ed. A. Ong and S.J. Collier, 439–463. Oxford: Blackwell.

Sass, H. 2008. Fritz Jahr's 1927 Concept of Bioethics. *Kennedy Institute of Ethics Journal.* 17 (4): 279–295.

Schrag, Z.M. 2010. *Ethical Imperialism: Institutional Review Boards and the Social Sciences, 1965–2009*. Baltimore: The Johns Hopkins University Press.

Van Den Hoonaard, W.C. 2011. *The Seduction of Ethics: Transforming the Social Sciences*. Toronto: University of Toronto Press.

Wainwright, S.P., C. Williams, M. Michael, B. Farsides, and A. Cribb. 2006. Ethical Boundary-Work in the Embryonic Stem Cell Laboratory. *Sociology of Health & Illness* 28 (6): 732–748. https://doi.org/10.1111/j.1467-9566.2006.00539.x.

Walters, L. 2003. The Birth and Youth of the Kennedy Institute of Ethics. In *The Story of Bioethics: from Seminal Works to Contemporary Explorations*, ed. J.K. Walter and E.P. Klein. Washington, DC: Georgetown University Press.

Wellmon, C. 2015. *Organizing Enlightenment: Information Overload and the Invention of the Modern Research University*. Baltimore: Johns Hopkins University Press.

Williams, C., J. Sandall, G. Lewando-Hundt, B. Heyman, K. Spencer, and R. Grellier. 2005. Women as Moral Pioneers? Experiences of First Trimester Antenatal Screening. *Social Science & Medicine* 61 (9): 1983–1992. https://doi.org/10.1016/j.socscimed.2005.04.004.

Elective Modernism and the Politics of (Bio)Ethical Expertise

Nathan Emmerich

1 Introduction

Whilst the question of ethical expertise is troubling in itself, more specific concerns can be raised when it is considered in relation to bioethics. Whilst bioethics is an inter- or, at least, multi- disciplinary field its dominant mode of thought is that of applied or practical ethics, understood as a particular form or mode of philosophical reasoning or as 'ethical rationality' in the analytic tradition. The claims or, at least, aims of this mode of thought suggest that right and wrong, good and bad, can be objectively determined through the unforced force of the better argument, to use a Habermasian turn of phrase. However, the very notion that there may be such a thing as expertise in applied or practical (bio)ethics[1] indicates that not all individuals are in the same position *vis-à-vis* 'the better argument.' It would seem that if (bio)ethical expertise is anything, then it involves knowledge of the bioethical literature coupled with a particular competence, ability or skill in the articulation, evaluation and adjudication of ethical arguments. If this is the case then it would seem that, were they to exist, experts in (bio)ethics would present a *prima facie* threat to the moral autonomy of individuals. If, say, some (bio)ethicists are experts in regards particular ethical questions that arise in the context of healthcare then it would seem that healthcare professionals ought to defer to them when encountering them in practice.

---

[1] In this essay I use the term (bio)ethics to mean the discussion and analysis of bioethical topics in accordance with the methodological prescriptions and philosophical presumptions of applied or practical ethics. The term bioethics denotes the broader inter- or multi-disciplinary field.

N. Emmerich (✉)
ANU Medical School, Australian National University, Canberra, Australia

The Institute of Ethics, Dublin City University, Dublin, Ireland

School of History, Anthropology, Politics and Philosophy,
Queen's University Belfast, Belfast, UK
e-mail: nathan.emmerich@anu.edu.au

Related concerns about the implications of (bio)ethical expertise arise in the context of democratic politics and policy-making. Such concerns are distinct from those that have been addressed to scientific expertise. Responses to the problem of scientific expertise either assert the value neutrality of science and the value-laden nature of politics or distinguish between scientific values and the other values that have a legitimate role to play in democratic debate and policy-making. In so doing the threat of technocracy – or epistocracy (Evans 2014) – can be countered whilst the independence and political autonomy of both science and democratic government can be maintained. However, the notion that there may be experts in ethics threatens to resurrect both the promise and the threat of technocracy. As Mari Levitt (2003) puts it 'what is the point of listening to the public when they have neither scientific nor ethical expertise?' Furthermore one might think similarly for any other group, such as politicians, who are similarly lacking in such expertise. Whether or not they are truly value-free, if scientists can provide the facts and ethicists the (broader) values why not embrace a technocratic approach to government?

In this essay my aim is to consider something of these broader questions and to do so in relation to the politics of the particular framework of expertise I have made use of in my previous analysis of (bio)ethical expertise (Emmerich 2015a, 2016). The conception of expertise I have previously worked with is the one Collins and Evans (2007) have pursued under Studies of Expertise and Experience (SEE) research programme and as part of what they call the third wave of science studies more generally. More recently Collins et al. (2010) have turned their attention to the political dimension of their work, arguing for what they call 'elective modernism.' Thus, the purpose of the below is to consider if, having adopted their theory of expertise, my account of (bio)ethical expertise can or should be associated with a similar political perspective or if the reorientation of expertise to the domain of (bio) ethics rather than science presents differing tensions in the socio-political exercise of expertise. In order to do so I first present a précis of my account of (bio)ethical expertise which is then followed by a brief summary of elective modernism. I then discuss whether elective modernism offers any insight into the socio-political uses of (bio)ethical expertise and its legitimacy.

2 A Socio-logical Account of (Bio)Ethical Expertise

The philosophical literature offers two different perspectives on ethical expertise and, albeit in their own specific manners, both raise the kind of concerns I gestured at above. Furthermore, as one is rooted within the discourse of modern moral philosophy (the contemporary account)[2] whilst the other is located with the (neo-)

---

[2]What I am calling the contemporary account is often rejected. Such rejection does not necessarily indicate a preference for the traditional account. Rather it involves a denial of ethical expertise *per se*. Those who make this move include individuals who could be seen as experts in ethics (Cowley 2005; Archard 2011). It seems to me that such rejections are primarily motivated by concerns about the (unethical or simply unpalatable) normative implications of ethical expertise, concerns that are, for the most part, a function of certain meta-ethical commitments. We should, instead, recognize that (bio)ethical expertise is a fact of contemporary society, modern cultural and democratic politics.

Aristotelian tradition (the traditional account), they not only significantly differ from one another, but would seem to be mutually incompatible. In addition both accounts are problematic. A significant failing of the contemporary account is the overly intellectualist nature of the view it presents. The contemporary account's adherence to 'the principle of phenomenalism' would seem to be incompatible with broader understandings of our (moral) psychology more generally (Narvaez and Lapsley 2005: 141). Whilst this could be taken to suggest that the contemporary account is on the right track – perhaps we should not expect ethical expertise to reflect our ordinary moral psychology – it does reflect the fact that it is inconsistent with broader research on the nature of expertise in general. Whilst some who work from within a neo-Aristotelian framework have developed the traditional account in the light of such expertise research (Stichter 2007; Swartwood 2013; Musschenga 2015) their view of ethical experts is primarily connected to the way some individuals – who we might call moral exemplars – lead their lives. In this view ethical expertise is less a matter an individual's cognitive, analytic or reflective abilities than it is a function of their virtues, their dispositions or, in a somewhat more sophisticated account, their 'social intelligence' (Snow 2009). Thus, whether it is considered in relation to their personal or their professional lives, what the traditional view offers is not something that reflects the behaviour or practices of all – or even most – (bio)ethicists or moral philosophers.

Adopting the framework of expertise – or *expertises* – developed by Collins and Evans (2007) I have addressed the conundrum of ethical expertise elsewhere (Emmerich 2015a, 2016). Briefly, we should distinguish between morality, and the *ubiquitous* expertise of all moral agents, and ethics, and the specialist ethics expertise of academic (bio)ethicists. The former is akin to philosophy's traditional account and, whilst it could be called a characterological or a virtue based theory, I prefer to talk of embodied dispositions and habitus. This is because if such expertise is ubiquitous then no claims about the moral standing of the bearers can be made or derived. The expertise lies in the agent's ability to negotiate the moral order or ethos of their socio-cultural (or political) context with ease. In so doing they may act in ways that are right or wrong, good or bad. Moral failings do not indicate a lack of expertise *per se* – despite a tendency towards rationalisation, post hoc justifications and a degree of 'denial' we are all too aware of our own moral failings. Thus ubiquitous moral experts need not be considered exemplars or moral saints. Rather they are associated with the ability to recognise and negotiate the moral landscapes we inhabit, something that both morally good and bad individuals are able to do.[3]

---

[3] Proof of this point can be found if one considers the philosophical literature on moral practice, where we find interesting sub-field in which the significance of psychopaths is discussed. Here the psychopath is held to be an individual who does not feel the compulsion towards morally good or right action – or away from morally bad or wrong actions – felt be the rest of the population. It is not that they are unaware of the moral landscape; even lacking any moral compunction, it seems they can negotiate it all too well. Thus psychopaths do not lack ubiquitous moral expertise; rather they lack the normative motivations (or 'conatus') required for them to act morally rather than immorally. Similarly consider the ubiquitous moral expertise of an individual situated in a socio-cultural or political context they consider morally abhorrent, but the structural dictates of which

It is worth noting a few additional features of this account. First, as implied, the moral order varies across different fields. Thus we find moral subcultures in which some are experts whilst others are not. Healthcare is an interesting example of one such subculture. Healthcare professionals have a ubiquitous moral expertise that relates to the respective practices they are engaged in. This is linked to, but differs from, their ordinary or everyday ubiquitous moral expertise. Second, whilst some try to differentiate them (cf. Smith 2009), the moral order or ethos of a field is, in essence, co-extensive with its normative order. Thus the notion of ubiquitous moral expertise includes practices usually differentiated from morality *per se*. Examples include etiquette, disciplinary norms and, indeed, any and all social structures that normative influence the social practices associated with a particular field. Both analysts and actors can, of course, focus on the moral order as a domain or sub-set of the normative order, and accord it a certain degree of priority or importance. Nevertheless, no formal or analytic distinction can be defended in theory or in principle; the moral and the normative are integrated within the concept of ubiquitous moral expertise. Finally, whilst ubiquitous moral expertise may include a reflective component – after all the exchange of reasons is central to what we might call our moral economy – it is primarily an unreflective phenomenon. Thus ubiquitous moral experts need not be able to justify themselves. Or, at least, they need only justify themselves according with the social norms of the contexts – or field(s) – they inhabit. There is, however, no need for them to be able to justify themselves in accordance with the norms of some other field, such as those we find in academic (bio)ethics.[4]

This conception of ubiquitous moral expertise contrasts with the specialist expertise of the academic (bio)ethicist. First, such expertise can be decomposed into two elements, what Collins and Evans call *contributory expertise* and *interactional expertise*. The former are those that can contribute to a particular discipline. In this case, and for simplicity's sake, this can be defined as those who write articles suitable for publication in the relevant journals. As all such individuals inhabit the field they also possess the ability to interact with their peers, and to do so in such a way so as to be recognised as members of the field. The nature of contributory expertise is such that it is predicated on this interactional ability.[5] However, some individuals may develop the relevant interactional expertise or, at least, some level of such expertise without becoming contributory experts. They have the ability to 'pass' in the field. Some science journalists are good examples of this phenomenon, as are

---

they are forced to obey. Thinking that such individuals lack the ubiquitous moral expertise required to do what is right, or that this disproves the notion of ubiquitous moral expertise, is to miss the point entirely. A large component of ubiquitous moral expertise is constitute by ones moral perceptions (Zahle 2013, 2014). Psychopaths can be understood as being possessed of such perceptual abilities whilst lacking any compulsion to follow its dictates.

[4]At least to some degree, these past two sentences account for the phenomena known as moral dumbfounding, see Emmerich (2016) for further discussion of this point.

[5]The reason being that the process of becoming a contributory expert involves being socialized (and enculturated) into the relevant field. This involves the development of interactional expertise See: (Collins and Evans 2015).

some sociologists of science.[6] Such individuals cannot 'walk the walk' but can, nevertheless, 'talk the talk.' Of course, given the notion of language as a practice (Collins 2011), talk is an indispensible component of scientific practices (and, one might add, the practice of bioethics). Thus, the ability to talk the talk can be understood as a matter of *walking the talk* (Collins and Evans 2007: chap. 4).[7] Therefore, properly conceived, a specialist contributory expertise involves the ability to walk the talk as an indispensible part of walking the (broader) walk, whereas the ability of specialist interactional experts resides in their ability to walk the talk alone.

The notion of interactional expertise is, I have suggested, of particular important for a proper understanding of (bio)ethical expertise. In the first instance, if they are to understand the fields and domains they comment upon, (bio)ethicists must develop some degree of interactional expertise with those fields and domains. In some instances this may be a fairly weak requirement, as when comprehending the biology of foetal development in order to comment upon the ethics of abortion, whilst in others it may be more demanding, as when commenting upon the ethical complexities of medical practice.[8] In the second instance, if we think that the point of bioethics is to influence those in other fields and domains, which is to say that if we think that the point of bioethics is to influence non-experts, then it would seem incumbent on the expert (bio)ethicist to develop the ability to effectively communicate with such individuals. Such an ability would draw on the ubiquitous moral expertise of both expert (bio)ethicists and those to whom they address their remarks.

Such an account of (bio)ethical expertise demonstrates the need for (bio)ethicists to think about the broader moral order or ethos of the fields they comment upon. However, one could still maintain that the substantive and methodological advice, recommendations and arguments offered by the academic field of applied or practical (bio)ethics ought to be embraced as they are objectively superior those found elsewhere. If we are to offset the kinds of concerns raised in the introduction, what is required is the addition of one science studies' most basic insights. Academic disciplines and scientific fields are (sub)cultures and, as such, they have moral orders, an ethos or a moral economy. Thus (epistemic) objectivity does not entail the absence of norms or values, rather objectivity "it is itself a code of values" (Daston and Galison 2007: 53). Scientific, indeed *academic*, disciplines entail that epistemology is fused with or wedded to an ethos (Daston 1995; Daston and Galison 2007: 204). This can be put another way. The generation of objective

---

[6] Part of the original impetus for the recognition of this sort of expertise was the abilities that one of the authors, Harry Collins, developed in relation to gravitational wave physics whilst conducting sociological research in this field (Collins and Evans 2007: 104–109).

[7] So as to avoid any potential misinterpretation that might result from the negative connotations usually attached to the notion of being able to talk the talk (whilst being unable to walk the walk) the phrase '*walk the talk*' is preferred by Collins and Evans (2007: chap. 4). It also makes clear that the 'talk' is very much part of the 'walk'.

[8] Some might think the order of these examples should be reversed. However, that is to underestimate the degree to which we already possess a certain level of interactional expertise with the field of healthcare. After all, we have all been patients. Therefore the interactional expertise required of (bio)ethicists builds on their wider, preexisting and non-academic experiences.

knowledge – or, indeed, any other type of knowledge – involves practices that are embedded within a specific social, cultural and historical context. Such knowledge is, therefore, necessarily dependant upon and informed by a particular normative or moral order. Neither science nor (bio)ethics is a value-free enterprise. They are both cultural phenomena.

3   Elective Modernism

What Collins and Evans have to offer, and what I have taken up, is a reconstruction of expertise. In their view this project is a necessity because, whilst wave 2 science studies has been highly successful in deconstructing science, we cannot do without the knowledge it provides. Therefore they propose a new direction for science studies – a third wave – one that they term Studies of Expertise and Experience. Whilst this is intended to compliment rather than replace Wave 2 research it is an attempt to move beyond what they see as the relatively naive impulse towards an ever increasing democratization of science. As such they aim to by provide "a set of tools for doing more than simply demanding 'more participation'" (Collins et al. 2010: 196). Their view is that we need to value, and balance between, democracy (populism) and expertise (technocracy). Elective Modernism is offered as one way in which a balance might be struck.

As such, elective modernism proposes that we (re)structure society by (re)organise the relationship between science and politics. It is a deceptively simple proposition that suggests we ought to "reconstruct the values of science" because "they are central to a good society" (Collins 2010). Rather than being seen as a resource – as, simply, a source of information or 'facts' – science is positioned as a key element of our contemporary – or modern – culture(s) (Collins and Evans 2007: 11), one that should be seen as worthy of our (political) respect.[9] Contra the perspective of 'second wave' science studies – where politics and science mix like wine and water – the third wave considers them immiscible or like oil and water (Collins et al. 2010: 194). This does not prevent science from being shot through by politics or with political concerns. Rather, it is to say that 'politics' is not part of what they call the 'formative intentions' of science. Whilst they acknowledge that the boundaries will always remain fuzzy, the notion 'formative intentions' is their thoroughly sociological approach to distinguishing between science and non- or pseudo- science.

[9] Democratic politics and policy making is said to be endangered by scientific expertise and 'scientism' more generally. Collins and Evans (2007: 11), defend a particular kind of scientism – scientism4 – that is summed up by the notion that science is an essential part of modern culture. Thus elective modernism is a form of scientism, but one that Collins and Evans defend. Similarly my work has been concerned by 'ethicism' as any articulation of (bio)ethical expertise must avoid the suggestion that we might abdicate our moral agency to a cadre of experts. The resolution I have adopted is analogous to Collins and Evans' notion of scientism4. It is to see bioethics as, in essence, one part of modernity's moral culture.

Formative intentions are the motivating social, cultural and – particularly in the case of science – epistemological structures that define a field, a mode of social life or, as they prefer it, a form of life.[10] They are the necessary – but not necessarily sufficient – components required if a particular mode of social life is to be possible. Thus, insofar as its funding must be determined via some political process and insofar as the relationship between different scientists will have a political component, politics remains a requirement for the contemporary existence of science. Nevertheless, politics does not form part of the fields formative intentions and is not, therefore, a component by *definition*. With this commitment Collins, Evans and Weinel found a distinction between science and politics and, on this basis, they can then engage in a normative discussion regarding the (re)configuration of their relationship. An important aspect of this is the way that they differentiate "between the technical and the political phases of technological decision-making in the public domain" (Collins et al. 2010: 186). Accordingly, whilst the political phase has priority or, at least, it has the final say in any decision- or policy-making process this does not mean it can seek to contrive, influence or otherwise "subvert the findings of the technical phase" (Collins et al. 2010: 188).[11] Similarly, contributions – and contributors – to the technical phase need to be carefully configured so that they do not cross over into the political phase.

Collins, Evans and Weinel suggest that the "technical phase is informed by the formative intentions associated with the scientific form-of-life, whereas the political phase is concerned with the formative intentions associated with the politics of the wider society" (Collins et al. 2010: 188). As such, when contributing to the technical phase of policy-making process scientists should speak as scientists and not on the basis of their religious, political or otherwise non-scientific beliefs. Thus, the technical phase does not refer to the pursuit of science itself, but to part of the political and policy-making process in which scientists offer their views or testimony. As such politically motivated interventions in the technical phase are not only illegitimate but political actors should not be considered free to distort, misrepresent or otherwise (re)interpret the information provided in this phase. Nevertheless, political actors remain free to ignore the technical input of science and scientists. Such a

---

[10] Whilst in my discussion of (bio)ethics and (bio)ethical expertise I did not make use the phrase 'formative intentions' the way in which I have sought to construe the field of (bio)ethics mirrors Collins, Evans and Weinel's understanding of science and scientific fields. Formative intentions have been equated with values (2010) as well as with ideals and vocabularies of motive (Collins et al. 2010: 191 & 198 note 10).

[11] Particularly when it is scientific research that has raised questions for policy-makers to address, it is clear that technical phase must, in some way, precede the political phase. However, when one considers specific cases and the process through which they are addressed in more detail it is not simply the case that one follows the other. When properly examined such decision-making processes are complex and move back and forth between their technical and political phases. Thus, the notion of a phase does not allude to their temporal sequence so much as to make metaphorical reference to their natures as being comparable different physical states, like gas or liquid (Collins and Evans 2007: 124 fn 17). The technical and political phases are, therefore, different social states, different contexts for ways of being or forms of life. Or, to my mind the better phrasing, different modes of social life.

view might strike some as an attempt to (re)institutionalize the distinction between fact and value that wave 2 science studies has collapsed. However, Collins, Evans and Weinel deny that this is the case and insist that they rely on the different formative intentions of different cultures – the ideological values and norms of any (sub)cultural domain or field – such as 'politics' and 'science.' This, the notion that "policy-makers should value the judgment of experts – those who 'know what they are talking about'" (Collins et al. 2010: 188) does not involve respecting science as, simply, a repository of facts or truth but as an important (sub)culture, and as an essential component of modernity. Indeed, as it involves *electing* to do so. It does not involve asserting the authority of science so much as adopting a particular political and evaluative stance with regard to that authority, its nature and limits, and its basis in the formative intentions of science and the values that underpin its practices.

To be clear, then, "what matters [for EM] is not that 'science', or scientific practice or scientific knowledge is chosen as the central element of our culture but that 'scientific values' are seen as being a key part of a democratic society" (Collins et al. 2010: 190). Indeed, consistent with the work of many others – including Merton and Habermas – the values of science are not only seen as being congruent with those of democracy, but as sharing overlapping values to a reasonably large degree (Collins et al. 2010: 191–192). Nevertheless, the 'formative intentions' of science and democracy differ; they are different fields, with different concerns and purposes. Thus, whilst electing to be modern is a political choice regarding the public and policy-making role of science, and the broader respect it is accorded, it is not only consistent with the (related and similarly modern) 'choice' to be democratic, and to value democracy, but can be represented as complimenting this broader commitment (Collins et al. 2010: 191). As such the socio-political role of science and scientific knowledge cannot be (re)configured at will. Respecting a field of inquiry means respecting its norms, its values and the formative intentions that constitute and underlie the 'community of practice.' As Collins and Evans say: "Democracy cannot dominate every domain – that would destroy expertise – and expertise cannot dominate every domain – that would destroy democracy" (2007: 8). A position that implies democracy and expertise should form a mutually supportive social compact.

What this means is that science can, after all, be demarcated from non- or pseudo-science but only in sociological terms. As such difficulties remain. Certain non-scientific fields may still resemble certain scientific fields. Nevertheless when we considered central examples such as, say, 'biology' and '(bio)ethics' the decision to term one a science and the other a non-science is fairly easy to make. Elective modernism suggests we can think similarly for the distinction between science and politics as a whole. This thinking can then be transferred when deciding who is in a position to speak as a scientist, an expert or as someone who 'knows what they are talking about' and who does not occupy such a position. The formative intentions that constitute the sociological distinction between different fields also indicate whether or not the Locus of Legitimate Interpretation (Collins and Evans 2007:

119–121) should be restricted to those positioned within the field or if it can be extended to others. For example, consider a field where there is little restriction on who may legitimately interpret is products, namely the artistic field. Individual artists – who inhabit particular social fields, the norms, formative intentions or ideologies of which they follow or, at least espouse – produce works of art. However, these works of art are available for interpretation by all and not just those who inhabit the artist's field. Aesthetic products – art, architecture, literature, fine foods and wines – can be legitimately interpreted by anyone.[12]

In contrast, the Locus of Legitimate Interpretation is far narrower when considered in relation to science. If one is to legitimately interpret scientific research one must be a contributory expert, or a very high level interactional expert. This is particularly true if one is to use such interpretation to conduct further scientific research. It also remains a reasonably strong condition for the interpretations that seek to (accurately) communicate scientific findings more generally. This restriction on legitimate interpretation is not, or so Collins (2010) suggests, the same as claiming that science is immune to criticisms leveled by non-scientists.[13] Clearly it can be subject to better or worse forms of criticism by, say, (bio)ethicists or Wave 2 sociologists. However, such criticism is rooted in alternative forms of expertise and, as such, it merely reinforces the duty of scientists to be clear when presenting and interpreting 'the science' to its wider audiences and publics.

Such a duty brings a final point into focus. It is rarely the case that the kind of interpretations required by non-scientists will involve the work of any one scientist. Rather, what is required are reports of the collective and consensus view of scientists and a scientific field. Whilst there are always disputes and disagreements, the technical phase of policy-making process involves scientists providing expert testimony as to the content and strength of the scientific consensus. Such expert testimony is not beyond *political criticism* but, under elective modernism, it is beyond *political (re)interpretation*. As such whilst those involved in policy-making process must respect the expert scientific interpretation(s) that have been provided during the technical phase, this is not to say that such testimony must acted upon in the political phase. Experts have to be heard, but they do not have to be obeyed (Evans 2014: 94). Nevertheless, where expert scientific advice is to be ignored or overruled there ought to be some acknowledgement of this fact. In short, "politicians must take responsibility for the policies they enact and be clear about the extent to which expert consensus supports these decisions" (Evans 2014: 94).

---

[12] Of course, this is not to deny that there might be elite, or even expert, consumers or interpreters of such works and products. Connoisseurship is acknowledged as form of (meta)expertise and, furthermore, it is a term that can be applied to science, to scientists and, in particular, to those involved in the practical and political management of science and scientists (Bourdieu 1996; Collins and Evans 2007: 57–59).

[13] Criticism by other scientists or those with scientific expertise such that they occupy the Locus of Legitimate Interpretation would, of course, not be an example of *criticism* but, rather, instances of further *interpretation*.

4   (Bio)Ethical Expertise and Elective Modernism

Before considering whether or not the academic field of (bio)ethics can be appropriately framed by the political perspective of elective modernism it is worth briefly
considering a couple of points regarding the relationship between science and (bio)
ethics, and the fact that (bio)ethics is in important channel for the public communication of science. It is also worth reiterating that my concern is with (bio)ethics as a
form of applied philosophical enquiry rather than bioethics more generally. This
latter term names the field as a whole and so includes disciplines – such as sociology, anthropology and history – that engage in broader and more critical forms of
analysis that are socially, culturally, and historically – and not just logically or, in a
somewhat limited sense of applied philosophy, ethically – reflexive. For these reasons I do not think bioethics as a whole could be properly considered in terms of
elective modernism.[14] Nevertheless, considering its relevance to the more limited
field of (bio)ethics will, I think, prove illuminating.

The relationship between science and (bio)ethics is an interesting one. Whilst the
ethical evaluation of scientific facts may vary – consider the differing interpretations
of the embryo's moral status – participants in such moral debates must have if not a
relatively undisputed and shared view of the facts then an electively modern one.
Whatever the stripe of their intellectual and, perhaps more importantly, ethico-
political perspective academic (bio)ethicists must relate to science in a way that
akin to elective modernism. This is not to say that (bio)ethicists may not legitimately disagree with the science or with a clinical perspective – some (bio)ethicists
may have a reasonably high level of interactional expertise and, therefore, will be
able to ask reasonably acute questions in the 'technical phase.' Nor is it to say that
they are necessarily complicit with the scientists in the way suggested by some
sociological accounts (cf. Evans 2012) although, of course, some might be.
Nevertheless, for the most part, (bio)ethics is committed to the distinction between
fact and value.[15] Whilst (bio)ethicists enact this distinction methodologically, it is
also present in the way such work is presented to its various publics, include scientists, politicians and all those involved in policy-making. Thus, whilst there is
(much) more to be said, elective modernism provides a potentially fruitful way to
frame the relationship between science and (bio)ethics. It also suggests that insofar
as (bio)ethicists are involved in the public communication of science – and, in my
view, they are heavily involved in this endeavour – then part of what they do,
communicating 'the science,' can be understood in terms of elective modernism.

---

[14] The difference is, of course, the degree to which different forms of intellectual enquiry respect –
or call into question – the ideological values of science and those that operate in practice.

[15] In a response to an article discussing elective modernism (Collins et al. 2010) Fischer (2011), a
sociologist of science, criticized the apparent revival of the fact-value distinction. However, as their
ongoing support for Wave 2 Science studies shows, Collins, Evans and Weinel are not seeking to
revive the fact-value distinction *per se*. Rather they are promoting the realization that the distinction between fact and value has, so to speak, *value*. Thus it may be adopted in some times, places
and contexts whilst rejected in others.

As interesting as these questions might be, they are not the particular concern of this essay. Rather, I wish to focus on the way the specific expertise of (bio)ethicists and the contribution it makes to policy-making processes might be understood in terms of elective modernism. In short, should non-(bio)ethicists take the expert advice – or 'testimony' – of (bio)ethicists in the same way as they do that of scientists? In my work on (bio)ethical expertise I have positioned applied philosophical (bio)ethics as a distinct social field and as an important aspect of our contemporary and modern moral culture. Such positioning reflects Collins et al.'s (2010) understanding of science and scientific fields. Similarly, (bio)ethical expertise is distinctive because of the formative intentions of this academic field; because of its socio-political and intellectual structure and the values that underlie and inform the research pursued and knowledge produced within the field. Furthermore, as is the case with science, when taken as a set the formative intentions of (bio)ethics, an academic discipline, are not to be found more generally. As such its products – journal articles – are not easily understood by the uninitiated. This is not to say that none of the values or norms of (bio)ethical discourse can be found outwith the academic field. Other fields can, of course, share certain values with (bio)ethics and, as is the case with science, (bio)ethics may have norms in common with the formative intentions of democracy. Nevertheless, the *formative* and *intended* ends of democracy, (bio)ethics and science differ. Such a view would, then, appear to suggest that Locus of Legitimate Interpretation of (bio)ethics compares to that of science rather than art, suggesting that might be restricted to experts.

However there is a sense in which one could say the same of literary criticism. As an academic field it is fairly inaccessible. Nevertheless, the existence of this intellectual field does not mean that non-scholars are unable to read, enjoy and interpret novels for themselves. Rather, the academic field and the field of literary consumption exist alongside one another and, on at least some occasions, interrelate with one another. For example, those who enjoy literature might be well advised to consider reading texts perceived by the academic field to be canonical, innovative or accomplished. Similarly, literary critics might be well advised to take note of popular culture and, in order to understand the value of popular works, its role, function and place in our literary cultures, to examine and engage with them as academics. Neither of these notions suggests that the views of experts in literary criticism are objectively superior to those of non-experts or, more accurately, those whose appreciation of literature is merely a function of ubiquitous expertise in the literary domain. Furthermore even if there may be something to gain for some, this does nothing to suggest that the ordinary reader has any particular need or use for expert literary criticism. Rather, it is merely to locate the field of literary criticism alongside, and as part of, our literary culture as a whole. This way of framing literary criticism is to reveal its relation to contemporary culture, a perspective that directly echoes the way in which elective modernism understands science as something related to and part of contemporary society and modern culture.

This same thinking can – indeed *must* – be transposed to (bio)ethics. The fact that (bio)ethics is an specialised field does not mean we should deny the legitimacy of medical doctors, life scientists or, indeed, lay persons interpreting their own moral experiences. Nor should we think that such interpretations should be subordinated to those of experts. Rather we need a more sophisticated account, one that positions the specialist expertise of (bio)ethicists as an important part of our broader moral culture and sees it in relation to the ubiquitous moral expertise of ordinary moral actors. This is part of what I address in my previous work. The existence of ubiquitous moral expertise and the nature of ordinary moral actors is such that we ought not blindly follow the advice, testimony or dictates of (bio)ethical experts. To do so would entail a significant abdication of our moral agency and, therefore, of our moral responsibilities. Nevertheless, if we are to be morally serious persons,[16] then we would be well advised to take note of what (bio)ethical experts have to say and to consider, criticise and interpret it for ourselves. This is particularly true for those who act (or practice) within fields like medicine. The ethical issues that arise within the contexts of modern medicine are not necessarily ones that individuals will be well prepared to address on the basis of their ubiquitous moral expertise or, at least, it is not simply the case that they are well prepared to do so. Whilst healthcare professionals – initiates to the field of medicine and healthcare – are morally socialized and ethically enculturated into this domain (Emmerich 2013, 2015b). The nature of this process, and of engaging in a relatively esoteric and specialised practice on an everyday basis, indicates the involvement and refinement or, better, contextual adjustment and development of the individual ubiquitous moral expertise. This is accomplished with the aid of the (bio)ethics, its literature and expertise.

This, then, indicates that there is a vital difference between ethics on the one hand and both science and aesthetics on the other when considered under conditions of elective modernism. Where the testimony of scientific experts can be criticised but not interpreted, and where the ordinary reader need not be concern themselves with the intellectual perspectives of literary critics, the case of (bio)ethical expertise differs.[17] Not only must it be criticised and interpreted by non-experts, the perspective developed and presented by expert (bio)ethicists should be consider of particular interest and concern to those who actually address and even encounter the issues

---

[16] The phrase is a subversion of Radcliffe-Richards (2012) more prescriptive and rhetorically loaded use of the same term. Albeit implicitly, Radcliffe-Richards' view would appear to suggest that all moral agents should become (bio)ethical experts, at least to the level of gaining significant interactional expertise with the field of applied ethics. At play here is a misguided assumption that extends what Narvaez and Lapsley (2005: 141) identify as the principle of phenomenalism – the notion that formal ethical reflection is a prerequisite for an act, behavior or practice to have 'moral significance.'

[17] It is, however, worth noting that it does not appear to apply to our ordinary or everyday ethical concerns but only to more specialist concerns of the kind raised by bioethics, business ethics, environmental ethics and so forth. We might ascribe this state of affairs to the way in which these domains require the careful evaluation of information, knowledge and perspectives that most are relatively uninformed about. A point that again highlights the role of bioethics in communicating 'the science.' Nevertheless it remains the case that, for the most part, applied ethics seems remarkably ill-suited to commenting on the ordinary ethics and moral practices of everyday life. A point that is, I would suggest, borne out by recent anthropological research (Zigon 2008; Lambek 2010; Faubion 2011; Laidlaw 2013). In a similar vein, see Johnson's (2014) remarkable and interdisciplinary 'Morality for Human Beings.'

(bio)ethics analyses. Furthermore, this should not be seen as a one-way street. Expert (bio)ethicists should concern themselves with the fact that ordinary moral agents who encounter (bio)ethical issues will take an interest in their work. Consistent with elective modernism, expert (bio)ethicists should, then, endeavour to make themselves and their work accessible to ordinary moral actors and, in so doing, contribute to broader discussions of bioethical issues.[18] Akin to the importance currently attached to the public communication of science we might think of the public understanding of (bio)ethics, with all that might be said to entail regarding raising awareness and the need to engage as well as communicate with broader audiences.

What, then, might this mean for democratic politics and policy-making processes? Clearly (bio)ethics – and bioethics more generally – should be viewed as an valuable part of our contemporary moral culture and the broader socio-political landscape. As such it has an important contribution to make with regards public debate, policy-making processes and, we might say, to the political life of a nation as a whole.[19] Equally, insofar as it has the potential to close down ethico-political debate and the full participation of non-experts, (bio)ethical expertise may present a threat to such democratic endeavours. Thus, (bio)ethicists ought to *take care* when exercising their expertise in arenas beyond their academic home. Interestingly, such an injunction is not only normative but, in essence, concerns the ethical or ethico-political limits that apply to the exercise of (bio)ethical expertise. If we first consider the use of (bio)ethical expertise in public debates then we might adopt certain standards of intellectual humility and generosity to our (expert and, in particular, non-expert) opponents. Furthermore, there is often a context-dependant case to be made for presenting a balanced opinion rather than, simply, advocating for a partisan position. Whilst the media often presents (bio)ethical debates in a 'for and against' format, the issues often require a more complex exposition if they are to be fully explored. Indeed, the 'for and against' format may itself be a source of imbalance and misrepresentation. Thus, if it is to be ethical, the exercise of (bio)ethical expertise may require individual experts to present a range of ethical perspectives or, at least, to intimate the degree to which the academic debate contains 'good faith' diversity.

This latter point has more acute relevance when it comes to the use of (bio)ethical expertise in political and policy-making processes. Whilst public debates can still be construed as ethical debates *per se*, this is not the case when we consider the more formal processes of policy-making. Whilst some might suggest that ethics should precede politics, that it should be understood as providing political discourse

---

[18] This does not, of course, imply that expert (bio)ethicists cannot engage in more esoteric, complex and expert forms of discourse. Just they, when required, they should at least make some attempt to communicate and engage with non-experts.

[19] Rosanvallon (2011) points out that politics and, indeed, policy-making, is not restricted to the work of the government but that, in modernity, has become 'decentered' with debates being distributed more widely. The UK's Nuffield Council on Bioethics is a good example of this decentering, and of the broader (bio)ethicsts make to 'the political life of a nation as a whole.'

with prior constraints or limits (Radcliffe-Richards 2012: 134). This is a view that implies that 'ethics' does not or should not occur or recur in subsequent discourses. I do not think this is a tenable position as, so to speak, 'the ethical' is inescapable; it is part of all aspects of our social and socio-political lives. Nevertheless the role of (bio)ethical experts is not to become partisans in such debate nor, worse, to structure them in accordance with academic methodologies.[20] Rather, on the basis of the fields cultural standing and formative intentions, (bio)ethicsts hold out the fields scholarship as a resource for political and policy-making debates, to (re)present the range of (bio)ethical perspectives and to do so both accurately and impartially. In so doing (bio)ethicists can contribute to the quality of political and policy-making discourses. Such notions reflect the way in which elective modernism construes the political and policy-making role of science and scientists. Thus, it would appear that despite not being in a position to testify to 'the facts,' the political functions of (bio) ethical expertise may bear significant comparisons with that of science. However, in order to maintain the notion that (bio)ethicists can contribute to political and policy-making process on the basis of their expertise we need to be able to distinguish between more than just expert and non-expert (bio)ethicists but between expert and *pseudo*-(bio)ethicists.

Whilst this iteration of the demarcation problem can, to some extent, be addressed in the way suggested by Collins and Evans, which is to say sociologically, problems remain. Consider, for example, the different academic, and therefore expert, approaches one can take to ethics. Whilst science is constituted by different fields – biology and chemistry for example – and whilst there is sometimes an overlapping focus – as, say, in the case of biochemistry – these fields are not in conflict. In contrast the substantive focus of different sub-fields of ethics not only overlap but are also in conflict over a range of substantive, methodological and meta-ethical issues. As such, whilst there may be such a thing as (bio)ethical expertise, and whilst there is some likelihood of there being a relatively widespread consensus – or, at least, a majority opinion – about the substantive position to take with regard a particular ethical issue, there is little to no chance that there will be agreement on the correct way to approach it (Toulmin 1981).

Of course, such things may or may not be taken as an indication that there is a fundamental problem with (bio)ethical expertise and/or its use within political and policy-making contexts. I tend not to think it overly problematic. Such disagreement is not restricted to the domain of academic or specialist (bio)ethical expertise

---

[20]This way of thinking often appears to be anathema to (bio)ethicists. Consider, for example, the ethical compromise on embryo research set out by the Warnock Report. From an applied (bio)ethical perspective the position adopted is rationally indefensible and it has been criticized by (bio) ethicists for this very failing (Harris 1985: 132). However, not only are there more nuanced views (Hammond-Browning 2015), consider the longevity and impact that the report has had on the regulatory landscape: it is, for example the basis of UK's Human Fertilization and Embryological Authority (HFEA) and, therefore, of the approval it recently granted for genome editing research. In regards its influence, durability and, more importantly, its political balance the report has been an outstanding success (Wilson 2011).

and any political process should be cognizant of the diversity of ethical perspectives that might bear on the substantive topic being addressed. Nevertheless there is, I think, a significant problem when it comes to the question of the degree to which we should prefer or prioritize the academic, specialist or specialized perspectives of (bio)ethical experts over those of ordinary moral agents, those who respond on the basis of their ubiquitous moral expertise alone. The existence of ubiquitous moral expertise, and the fact that specialist (bio)ethical expertise could not be developed without it, undermines a key aspect of the demarcation problem. Lest we forget the demarcation problem concerns the distinction between science and non- or pseudo-science. Whilst those working in science studies need worry about certain forms of lay expertise – Cumbrian sheep-farmers, say – their expertise is not a claim to scientific knowledge *per se*, but to knowledge based on experience. In contrast the views of ubiquitous moral experts are *ethical* views. Whilst we can maintain a sociological distinction between specialist (bio)ethical expertise and ubiquitous moral expertise, the fact that they both belong to the same class – morality – is one of the reasons that the testimony that specialist (bio)ethical experts provide in the technical phase cannot be considered beyond reinterpretation in the political phase. This view is reinforced by the principle of (moral) equality, understood as a formative intention or value shared by democracy and (many) moral and political philosophies. Unless one has reason to think that they are held in 'bad faith' the formative intentions or values of modern society indicate that the views of all moral agents are deserving of our consideration and respect. Of course, that a particular moral view is held by some and that it therefore deserves consideration may not entail very much at all, and it certainly does not prevent us from engaging with it and those who hold it in further ethical and ethico-political debate. Nevertheless, one must conclude that there is no such thing as pseudo-ethics or pseudo-ethical perspectives.

Such concerns prevent us from committing to elective modernism when it comes to (bio)ethical expertise. Whilst (bio)ethics can be seen as an aspect of our moral culture, and not simply a resource, and whilst it can be appropriate to take particular note of the views held by (bio)ethicists in virtue of their specialist expertise, these cannot be considered to be beyond the (re)interpretation of other moral agents, even if they can only do so on the basis of their ubiquitous moral expertise. Furthermore, politics and policy-making should themselves be seen as moral endeavours. Indeed, given that, unlike science, politics and (bio)ethics are highly miscible – that it is very easy to slip from doing ethics into doing politics – we must actively work to ensure the political neutrality (or 'objectivity') of specialist (bio)ethical expertise. This is task that should weigh heavily on the shoulders of the expert (bio)ethicist and, ideally at least, on politicians. What this view suggests is, I think, that elective modernism is a political philosophy with particular relevance to contemporary relationship between science and society. However, once we configure our democratic processes to reflect this relationship it will have consequences in other areas, such as in the case of (bio)ethics. At the heart of elective modernism is a choice about the role of science and the relevance of the fact-value distinction to that role. Following this choice means that the distinction becomes procedurally institutionalised as the technical phase and the political phase. Insofar as (bio)ethicsts adopt the same distinction, then they appear to fit well into the same schema. Nevertheless the notion that elective modernism can correctly frame the political contributions of (bio)ethical experts cannot be maintained as what they have to contribute are not matters of fact but ethical perspectives and matters of value, and, in the final analysis, there can be no (expert) value – (lay) value distinction.

5   Conclusion

This essay has pursued the notion of (bio)ethical expertise and elective modernism with a view to their compatibility. The conclusion I have drawn is that in describing the relationship between science and society and its political role, elective modernism has a significant influence on how we might understand the role of (bio)ethical expertise. Nevertheless, due to the fact that values are the focus of its enquiry the relationship between (bio)ethics and society is incompletely captured by elective modernism. Its influence might, then, be traced to the important role that science has to play in informing (bio)ethical discourse and, in turn, the role (bio)ethics plays in the broader communication of scientific perspectives. Somewhat ironically then, science and (bio)ethics are entangled by the very distinction that keeps them apart; both are inescapable related insofar as the formative intentions of both fields involves a methodological commitment to the independence of fact and value. However, where democratic politics can elect to respect the independence of facts it cannot indeed should not, do the same for values. This is because our values are interdependent, a notion that is present in the idea that science and democracy share certain values. Similar implications can be sketched with regards bioethics as a whole. Whilst (bio)ethics rarely concerns itself with democratic principles as part of its substantive analysis, this is not the sum total of the discipline's endeavours. For example, Montgomery (2013, 2016) has recently argued for the value of public bioethics and for understanding bioethics as a governance practice.[21]

Such thinking points to a specific limitation in the way I have restricted my analysis of bioethical expertise to (bio)ethics. There are very few individuals who restrict themselves to (bio)ethics and, furthermore, much of the way I have discussed the ethics of (bio)ethical expertise implies that such experts should go beyond the practice of (bio)ethics alone. What I take this to mean is that the cultural tasks and socio-political roles fulfilled by bioethics are diverse, a fact that reflects the morally plural context in which it takes place and to which it must respond. Contra to what Radcliffe-Richards (2012) recommends, moral seriousness is not comprised of settling frameworks for ethical debates prior to the conduct of any concrete policy-making process. In short not only is (bio)ethical expertise not the sum total of morality neither is it comprised of (bio)ethical expertise plus some

---

[21] Of course the fact that, at the time of writing, Montgomery was Chair of the Nuffield Council for Bioethics is pertinent to his view, and vice versa.

degree of political modulation via the views of the 'lay' public. In the context of contemporary or modern democracies, ethico-political decision-making ought to be seen as a complex, decentred, distributed and reflexive (Rosanvallon 2011) process that will never be entirely complete.

References

Archard, D. 2011. Why Moral Philosophers Are Not And Should Not Be Moral Experts. *Bioethics* 25 (3): 119–127. https://doi.org/10.1111/j.1467-8519.2009.01748.x.

Bourdieu, P. 1996. *The Rules of Art: Genesis and Structure of the Literary Field*. Stanford: Stanford University Press.

Collins, H. 2010. Elective Modernism. http://www.cardiff.ac.uk/socsi/contactsandpeople/harry-collins/expertise-project/elective%20modernism%204.doc. Accessed 1 Feb 2016.

———. 2011. Language and Practice. *Social Studies of Science* 41 (2): 271–300. https://doi.org/10.1177/0306312711399665.

Collins, H., and R. Evans. 2007. *Rethinking Expertise*. Chicago: University of Chicago Press.

———. 2015. Expertise Revisited, Part I – Interactional Expertise. *Studies in History and Philosophy of Science Part A* 54: 113–123. https://doi.org/10.1016/j.shpsa.2015.07.004.

Collins, H., M. Weinel, and R. Evans. 2010. The Politics and Policy of the Third Wave: New Technologies and Society. *Critical Policy Studies* 4 (2): 185–201. https://doi.org/10.1080/19460171.2010.490642.

Cowley, C. 2005. A New Rejection of Moral Expertise. *Medicine, Health Care and Philosophy* 8 (3): 273–279. https://doi.org/10.1007/s11019-005-1588-x.

Daston, L. 1995. The Moral Economy of Science. *Osiris* 10: 2–24.

Daston, L., and P. Galison. 2007. *Objectivity*. New York: Zone Books.

Emmerich, N. 2013. *Medical Ethics Education: An Interdisciplinary and Social Theoretical Perspective*. London: Springer.

———. 2015a. A Sociological Analysis of Ethical Expertise: The Case of Medical Ethics. *SAGE Open* 5 (2): 1–14. https://doi.org/10.1177/2158244015590445.

———. 2015b. Bourdieu's Collective Enterprise of Inculcation: The Moral Socialisation and Ethical Enculturation of Medical Students. *British Journal of Sociology of Education* 36 (7): 1054–1072.

———. 2016. A Sociological Analysis of Ethical Expertise: The Case of Bioethics. *Cogent Social Sciences* 2 (1): 1143599. https://doi.org/10.1080/23311886.2016.1143599.

Evans, J.H. 2012. *The History and Future of Bioethics: A Sociological View*. New York: Oxford University Press.

Evans, R. 2014. Science and Democracy in the Third Wave Elective modernism Not Epistocracy. In *Expertise and Democracy*, ed. C. Holst, 85–102. Oslo: ARENA, Centre for European Studies, University of Oslo https://sv.uio.no/arena/english/research/publications/arena-publications/reports/2014/report-01-14.pdf#page=92. Accessed 21 Oct 2014.

Faubion, J.D. 2011. *An Anthropology of Ethics*. Cambridge: Cambridge University Press.

Fischer, F. 2011. The "Policy Turn" in the Third Wave: Return to the Fact–Value Dichotomy? *Critical Policy Studies* 5 (3): 311–316. https://doi.org/10.1080/19460171.2011.606304.

Hammond-Browning, N. 2015. Ethics, Embryos, and Evidence: A Look Back at Warnock. *Medical Law Review* 23 (4): 588–619. https://doi.org/10.1093/medlaw/fwv028.

Harris, J. 1985. *The value of life*. London: Routledge.

Johnson, M. 2014. *Morality for Humans: Ethical Understanding from the Perspective of Cognitive Science*. Chicago: University of Chicago Press.

Laidlaw, J. 2013. *The Subject of Virtue: An Anthropology of Ethics and Freedom*. Cambridge: Cambridge University Press.

Lambek, M. 2010. *Ordinary Ethics: Anthropology, Language, and Action*. New York: Fordham University Press.

Levitt, M. 2003. Public Consultation in Bioethics. What's the Point of Asking the Public When They Have Neither Scientific nor Ethical Expertise? *Health Care Analysis* 11 (1): 15–25.

Montgomery, J. 2013. Reflections on the Nature of Public Ethics. *Cambridge Quarterly of Healthcare Ethics* 22 (1): 9–21. https://doi.org/10.1017/S0963180112000345.

———. 2016. Bioethics as a Governance Practice. *Health Care Analysis* 24 (1): 3–23. https://doi.org/10.1007/s10728-015-0310-2.

Musschenga, A.W. 2015. Editorial: VIRTUAL ISSUE No. 1: Virtues, Skills, and Moral Expertise. *Ethical Theory and Moral Practice,* (Virtual Issue, Online Only). http://www.springer.com/social+sciences/applied+ethics/journal/10677. Accessed 2 Feb 2015.

Narvaez, D., and D.K. Lapsley. 2005. The psychological foundations of everyday morality and moral expertise. In *Character Psychology and Character Education*, ed. D.K. Lapsley and F.C. Power, 140–165. Notre Dame: University of Notre Dame Press.

Radcliffe-Richards, J. 2012. *The Ethics of Transplants: Why Careless Thought Costs Lives*. Oxford: Oxford University Press.

Rosanvallon, P. 2011. *Democratic Legitimacy: Impartiality, Reflexivity, Proximity*. Trans. A. Goldhammer. Princeton: Princeton University Press.

Smith, C. 2009. *Moral, Believing Animals: Human Personhood and Culture*. Oxford: Oxford University Press.

Snow, N.E. 2009. *Virtue as Social Intelligence: An Empirically Grounded Theory*. London: Routledge.

Stichter, M. 2007. Ethical Expertise: The Skill Model of Virtue. *Ethical Theory and Moral Practice* 10 (2): 183–194. https://doi.org/10.1007/s10677-006-9054-2.

Swartwood. 2013. Wisdom as an Expert Skill. *Ethical Theory and Moral Practice* 16 (3): 511–528.

Toulmin, S.E. 1981. The tyranny of principles. *The Hastings Centre Report* 11 (6): 31–39.

Wilson, D. 2011. Creating the "Ethics Industry": Mary Warnock, In Vitro Fertilization, and the History of Bioethics in Britain. *BioSocieties* 6 (2): 121–141.

Zahle, J. 2013. Practices and the Direct Perception of Normative States: Part I. *Philosophy of the Social Sciences* 43 (4): 493–518. https://doi.org/10.1177/0048393112454995.

———. 2014. Practices and the Direct Perception of Normative States: Part II. *Philosophy of the Social Sciences* 44 (1): 74–85. https://doi.org/10.1177/0048393112462517.

Zigon, J. 2008. *Morality: An Anthropological Perspective*. Oxford: Berg.

Grounding Knowledge and Normative Valuation in Agent-Based Action and Scientific Commitment

Catherine Kendig

1 Introduction

If the goal of scientists is the acquisition of knowledge, then knowledge in general and scientific explanation in particular can surely be understood to be the product of that pursuit articulated as a unified set of observation statements. However, those focusing on scientific practice, disagree. They claim that knowledge is always understood with reference to a particular context and in light of the actions of an epistemic agent. Knowledge-making activities are not the result of universal rules for deriving explanation from facts but the result of critical intersubjective modes of investigation. A science-in-practice approach turns our attention to the activities of and communication between scientists in order to understand and characterize the nature of scientific inquiry. As such, it is part of what has been referred to as *the practice turn*.[1] This refocusing of science on scientific practices highlights the activities that are revealed when we look at the processes and doings of science by scientists and scientific communities (e.g. hypothesizing, testing, experimenting, theorizing, measuring) rather than exclusively on the products of science (e.g. knowledge, equations, devices, theories). The practice turn in philosophy of science is not an apologetic for an and-practice-too approach to the metaphysics of science. A focus on practice provides a route to understanding the nature of the world in ways that have been, until recently, marginalized by aggressive demarcationist interests within traditional philosophy of science. This form of aggressive demarcationism held that research seeking to investigate scientific activity and the work of scientists was *not really* philosophy *but*

---

[1] For research within the philosophy of science in practice and sociology of science in practice, see for instance the work of Hacking (1992, 1995), Dupré (1993), Chang (2004), Rouse (1996, 2003), Rheinberger (2005), De Regt et al. (2009), Soler (2012), Soler et al. (2014), and Kendig (2016b, c).

C. Kendig (✉)
Department of Philosophy, Michigan State University,
503 South Kedzie Hall, 368 Farm Lane, East Lansing 48824-1032, MI, USA
e-mail: kendig@msu.edu

© Springer International Publishing AG, part of Springer Nature 2018   41
H. Riesch et al. (eds.), *Philosophies and Sociologies of Bioethics*,
https://doi.org/10.1007/978-3-319-92738-1_3

just sociology.[2] Within philosophy, there was no worse criticism than the suggestion that what one's putatively philosophical research was up to was not philosophy but was instead sociology. For many in philosophy, the criticism amounted to disciplinary slur, and one to be avoided. This active avoidance by philosophers of science to engage with sociology of science meant that rather than crossing the divides between philosophy and sociology, many were instead burning any bridges that remained between them in an effort to protect the discipline from invasion.

Rather than seeking a purely theoretic approach to knowledge or wholly analytic approach to explanation, a philosophy of science in practice approach focuses on the activities required in theory-making, knowledge-making, and explaining. Why is this important? Focusing on an activity-based analysis allows us to "go beyond thinking about scientific explanation in terms of logical relations between explanandum and explanans, [and] we can consider how the act of explaining arises and how it is best performed" (Chang 2011: 208). So called pure theoretic approaches that omit reference to practice succeed in doing so only by assuming science and knowledge acquisition to be a subjectless state of affairs—activities with no actors, understanding with no one who understands, and modelling with no modellers. Chang suggests the solution to this problem is for us to go against the convention of avoiding the second person familiar "you" in our discourse, explanations, and discussions (Chang 2011). We should (as philosophers, sociologists, and scientists) recover the importance of what knowledge is as something you or I understand or explain, rather than as disembodied subjectless answers to questions (Chang 2011, 2016).

Implementing a science-in-practice approach, my aim is to turn attention to the work of practitioners reengineering metabolic pathways within chassis organisms such as *E. coli*. I ask, what, if anything, doing so can tell us about the relationship between the metaphysical, epistemological and ethical knowledge-making activities. As such it constitutes an activity-based analysis of scientific explaining and normative ethical thinking. If successful, it would suggest that an examination into the practice of science may also provide answers, (or at least more informed lines of questioning), for other long-discussed problems in philosophy.

I begin with a brief metaphysics and epistemology of classification. Disciplines have a system of classification that specifies the kinds of things that are the subject of study for that discipline, e.g. the periodic table of elements, plate tectonics, the DSM (Dupré 2006). Synthetic biology is no different. To understand the classification system one must focus on the processes by which it is used and made. Our behaviour is informed by what kinds of things we (presuppose) we are interacting with as well as the goals and values we rely upon in our investigation. In the first half of the chapter, I examine the nature of scientific inquiry and how the manipula-

---

[2] Chang has also pointed out the tendency of traditional philosophers of science as well as analytic philosophy in general to use the "just sociology" claim as criticism of practice based approaches to philosophy: "In the typical analytic philosopher's picture, the scientist only enters as a ghostly being that either believes or doesn't believe certain descriptive statements, fixing his beliefs following some rules of rational thinking that remove any need for real judgment. All the things that do not fit easily into this bizarre and impoverished picture are denigrated as pieces of "mere" psychology or sociology" (Chang 2014: 70).

tion and use of different techniques within synthetic biology has metaphysical implications for the notion of parts and wholes, modularity, biological organization and function. The second half explores the social aspect of scientific inquiry in an attempt to reveal how normative valuations and ethical judgements are formed within and across synthetic biology communities. It concludes with a suggestion for how the epistemological, metaphysical, and ethical modes of inquiry are connected in networks of practitioners working together as moral agents—and that kinds of moral objects, epistemic objects, and metaphysical objects are made kinds through the activities of practitioners within these networks.

2 Categories of Epistemological Activity in Synthetic Biology

Early conceptual work in synthetic biology identified three different knowledge-making distinctions that exist within the field (see O'Malley et al. 2008; O'Malley 2009; Morange 2009a). These were intended to distinguish three overlapping epistemological categories. The categories demarcate diverse knowledge-making distinctions that lead to different questions being asked, different methods used, different knowledge acquired through these, and different products or outcomes (O'Malley et al. 2008). The first of these three categories is whole genome engineering. In this, biological processes and modules are co-opted and redesigned to solve technological problems for the production of energy or chemicals required for various industries. The most publicized example of this was the synthesis of the first self-replicating, synthetic bacterial cell. This was achieved by Craig Venter and his company, Synthetic Genomics, who synthesized an entire bacterial genome and used it to replace the genome of *Mycoplasma mycoides* thereby creating JCVI-syn1.0 (Gibson et al. 2010; Hylton 2012). The second is the engineered construction of functional parts, processes, pathways, devices, and systems (Brent 2004; Endy 2005). The current attempts to modify metabolic pathways in bacteria, yeast, and algae to generate biofuels are examples of this (Dellomonaco et al. 2010; Georgianna and Mayfield 2012; Wang et al. 2013). This involves the design and testing of biological systems and their component parts. Understanding of these functional systems is born out in their decomposition, manipulation, and co-option. Understanding of these parts and networks is based on the structure and syntax of the system. Describing the syntactic structure of the biological network using engineering, logic, or mathematical models, researchers can gain knowledge of the behaviour of the module or pathway in terms of its input and output conditions. The third category of investigation is synthetic experimental evolution or protocell creation (Erwin and Davidson 2009; Morange 2009a). In this endeavour, synthetic biology seeks to understand the process of evolution, biological organization, and the nature of modularity. Understanding the process also opens up the potential to optimize the modules, networks, and systems which direct it.

This initial tripartite classification of synthetic biology provided important knowledge-making distinctions in the modes of research and knowledge based on these but it also led to further metaphysical questions. For instance, what *kinds* are the parts and processes to which these categories refer? And, what kinds of things are the parts, modules, and systems used within the discipline of synthetic biology? Pablo Schyfter[3] (2012) suggests what he calls an initial "exploration into things and kinds" offering a "first look analysis" of kindhood for the products of synthetic biology and suggests that they fit imperfectly within both technological kinds and natural biological kinds. He is critical of synthetic biologists for not considering kindhood and for using engineering as the model and exemplar on which to base synthetic biology.

Determining what kinds (i.e technological kinds or natural biological kinds or something else) exist in synthetic biology may ultimately rest on how the discipline itself and how it is understood in relation to other biological and engineering disciplines. The discipline of synthetic biology is sometimes conceived of by practitioners and detractors as a subset of functional biology and as such is characterized as an application-based, or technology-based mode of understanding that seeks to explain how something works (see Schyfter 2012 for problems with this view). It has also been characterized as evolutionary biology due to its attempt (especially in protocell creation) to answer why-questions: seeking why (rather than how) biological pathways, devices, and parts work. This difference in the attribution of goals, products, and techniques depending on what types of questions are being asked make the categorization of synthetic biology as a hybrid or disunified discipline unsurprising. Its growing epistemic and methodological toolkit seems likely to continue apace-- the result of sourcing and modifying techniques from biology, chemistry, computer science, mathematics, and engineering (see Morange 2009b; Keller 2009; O'Malley et al. 2008 for discussions of the discipline-building of synthetic biology).

But relying on this dichotomy of functional and evolutionary biology, of how- and why-questions does not seem entirely justified—or at least is not always elucidatory—within synthetic biology. Knowledge-seeking questions within synthetic biology do not focus purely on how-questions directed for the purpose of modifying function. I suggest elsewhere (Kendig 2016a, b) that such dichotomizations fail to identify the union of how- and why-questions, their mode of investigation, and categorization and kind-making (or what I've called "kinding") practices typified by synthetic biology. The making or constructing of material objects, mechanisms, processes, or pathways; the theoretical construction of models and algorithms; as well as the devising of repeatable methods and techniques being made in synthetic biology are all instances of kinding—where kinding is understood as the epistemological and ontological activities within the practice of synthetic biology and by which the categories of that discipline or subdiscipline are configured. The outputs of these practices—the products of diverse synthetic biological research aims, are exchangeable and repeatable activities that represent, explain, and further advance

---

[3] Schyfter (2012) considers and evaluates the appropriateness of conceiving the products of synthetic biological research as kinds of technological objects. The discussion here differs from his insofar as I take a practice-based account of kinds that focuses on kinds of modules (see also Sprinzak and Elowitz 2005; Keller 2009).

our understanding of the relation of parts and wholes, the manipulation of developmental pathways, and the nature of biological functioning and organization. This is exemplified in the engineered construction of functional parts, processes, pathways, devices, and systems (Brent 2004; Endy 2005). For example, the current attempts to modify metabolic pathways in bacteria, yeast, and algae to generate biofuels (Dellomonaco et al. 2010; Georgianna and Mayfield 2012; Wang et al. 2013) rely on the modification of metabolic pathways through design and testing of biological systems and their component parts.

Understanding of these functional systems is born out in their decomposition, manipulation, and co-option. The type of synthetic biology focused on the engineered construction of functional parts, processes, pathways, devices, and systems is in the business of producing standardized parts, devices, pathways and modules with known functions. Standard biological parts with known functions are catalogued in a number of registries (e.g. Massachusetts Institute of Technology Registry of Standard Biological Parts). Insofar as these parts are kinds, this practice is the making of these parts.

In response to Schyfter (2012), I suggest that the imperfect fit of the parts and processes of synthetic biology into technological or biological kinds is because that which is kinded is epistemologically heterogeneous. If synthetic biology provides knowledge of how systems work, then the explanations of nature it provides (e.g. about how to re-engineer and manipulate them) suggest that it may be more profitably conceived of as a discipline that is epistemologically, ontologically, and methodologically hybrid. In the next section, I characterize the practice of part-making in metabolic engineering. I suggest that this practice can be couched in terms of different kinds of modularizing. It can be understood as a study of kinds of synthetic biological objects in the making and the nature of those things that make up the discipline (and subdisciplines) of synthetic biology.

3   The Tangle of Modularity

The activities of practitioners whose work focuses on the engineered construction of functional parts, processes, pathways, devices, and systems appear to be at least in part based on an underlying philosophical premise—that the world is organized in a certain way—that living organisms have an organization that is modular and that the variation that natural selection acts upon results from the recombining of modules. The premise of biological modularity is an ontological claim that appears to come out of this particular form of synthetic biological practice. We understand that the biological world is modular because we can manipulate different parts of organisms in ways that would only work if there were discrete parts that were interchangeable. This is the foundation of the BioBrick assembly method widely used in synthetic biology (Knight 2003). It is one of a number of methods that allows practitioners to construct and reconstruct biological pathways and devices using DNA libraries of standardized parts with known functions.

Because of its pick-and-mix approach to disciplinary tool sourcing, synthetic biology may be particularly well-suited to unpick the tangled meanings of concepts that have been long-debated within evolutionary and functional biology (Morange 2009a: 374). *Modularity, constraints,* and *convergence* can be stripped down and analysed and given functional parts-language descriptions through synthetic biological experimentation and manipulation of genomes of model organisms, such as that of the synthetic biology workhorse, *E. coli.* In the following, I explore how metabolic engineering, as an epistemic tool, may be used to disentangle the multiple notions of two of these concepts—*modularity* and *fitness.*

Modularity has been a concept much discussed (Schlosser and Wagner 2004; Wagner et al. 2007). Although seen to be a central principle of functional biology in general and synthetic biology in particular, it continues to cause confusion. Within the emerging discipline of synthetic biology, this inbuilt ambiguity means that context is usually required for the use of the term to confer knowledge about the system from one practitioner to another.

"Modularity" is used to identify at least two different kinds of claims within synthetic biology. I refer to these different forms following the philosophical convention for discriminating broad claims from specific claims of concepts using uppercase and lowercase letters: *Modularity* and *modularity.*[4]

1. *Modularity.* A general thesis about the nature of all biological organization as being made up of relatively independent components, like building blocks. This can be a metaphysical claim: that the organization of living things is Modular; a claim that applies with univocality to all biological structures. It can be a methodological claim; that the best way to proceed with research is to look for Modularity. It can be an epistemic claim, that the world is knowable and we gain understanding by considering it is Modularly organized.[5]

2. *modularity.* A specific claim about the parthood of a particular organism, pathway, device, or process. This conception focuses on the property of being a part and the metaphysical relationship of individuation or composition of it as it relates to a whole. The claim of modularity is one about the nature of parts and wholes. Included under this kind of claim are various sub-characterizations, for instance: that parts are relatively autonomous, interchangeable, or independent of the context of other parts, carry out measurable functions, standardisable, or are non-decomposable.

These two forms frame different spaces of epistemological and ontological investigation. In doing so, they configure the level of biological organization to which the

---

[4]When referring simultaneously to "Modularity" and "modularity" I will use the admittedly awkward "M/modularity".

[5]Each of these claims can be either conceived of from a realist, antirealist, operationalist, or pragmatic view as well as either one of monism or pluralism. For instance, one might suggest Modularity is a pragmatic methodology (and that we can be agnostic about whether the world is or is not really Modular). Someone may justify this claim that it is Modular insofar as our best knowledge comes from a working hypothesis of Modularity that is a heuristic guiding synthetic biology research.

theses of M/modularity apply, and the referent to which they apply (e.g. processes, objects, relationships, and properties).

3.1 Assembly Methods Affecting Modularity

Synthetic biology is based on a general thesis of biological Modularity (Knight 2003; Morange 2009a). The parts database of biological modules with well-characterized functions relies on this working hypothesis and the success of synthetically produced networks and parts in practice seems to bear this out at least in some limited capacity. Taking a practice-focused approach to understanding Modularity suggests that the assembly method one uses affects the nature of Modularity of the parts, devices, and pathways one constructs.

The most commonly used assembly method, BioBrick assembly, is based on the standardization of basic parts with a prefix and a suffix (Knight 2003). Addition of these two short DNA sequences to any DNA element results in a BioBrick. Two BioBricks can be ligated together in either order to construct a composite that is itself a BioBrick containing a prefix and a suffix (Kendig and Eckdahl 2017). The central concept that enables BioBrick assembly to work is that type II restriction enzymes XbaI and SpeI produce compatible sticky ends (Kendig and Eckdahl 2017). BioBrick assembly has been a popular method from its inception. However, it is not without problems. BioBrick assembly depends on purification of DNA fragments after gel electrophoresis. The pairing of the DNA recognition sequences for XbaI and SpeI during ligation results in a mixed site that practitioners refer to as a BioBrick scar (Brent 2004; Endy 2005). The scar is six DNA nucleotides in length and has a defined sequence. There are applications in which the spacing between two parts must be smaller than six nucleotides or when the BioBrick scar sequence confounds the function of a device (Kendig and Eckdahl 2017). Another problem with BioBrick assembly is that once ligated together, two BioBricks cannot be taken apart for use in another construct (Knight 2003). Assembled BioBricks are components that can no longer be swappable or interchangeable. In this way, the generation of the scar during BioBrick assembly affects the kinds of parts produced and their degree of Modularity. Being Modular means something different during BioBrick assembly because of this known outcome.[6]

3.2 Modularity and Modularizing

The claim of modularity is a claim specific to the compositional structure that parts and wholes exemplify in a particular system. As such, modularity is an answer to questions concerning: What is the nature of parts in that particular device or

---

[6]An extended discussion of modularity based on this example is contained in Kendig and Eckdahl (2017).

pathway?, How are they organized?, or What is the relationship of the compositional structure of them to the whole in question? It characterizes the property and
nature of modulehood. This second kind of modularity avoids the problems that
befall Modularity regarding different assembly methods. The things that are catego-
rized as modular are numerous and may come about due to differing processes of
assembly or categorization. This kind of modularity defines modules by the action
of modularizing. The assembly method used makes parts into modules insofar as
they are chosen by practitioners to be connected or separated from other parts. What
counts as a module (variously understood) and the criteria for modular kindhood
may not be the same for all modules. That is, modularity may not be a property that
can be univocally expressed for all parts—modularity allows for this.

This kind of modularity is not without problems. To say something is a module
or part has typically meant that it bears some sameness relationship, family resem-
blance, or overlapping shared homeostatic set of properties to another part qua
module. In this way, being a module—insofar as it is understood to be a property—
means that it is a property that is instantiable in one way. This means that all modules
insofar as they are modules are homogeneously so. I think this is a mistake. This
modulehood may be a kind differently instantiated. That is, modulehood for one
thing may not be the same as modulehood for another in a radical and non-
comparative way. Some might suggest that this brings into question the legitimacy
of what it is that we refer to when we claim something is a module if there is no
unifying claim. When faced with this proposition, they may prefer to dispense with
all claims of modularity and become eliminitivists. Alternatively, they may embrace
the heterogeneity of modulehood and allow that modulehood may be radically
heterogeneous across all metaphysically parcelled out stuff. Despite its initial
simple understanding, all modules may not belong to the same sui generis category—
modularity may not be something that is univocally expressed.

4   Evolvability

I move now from mapping out some of the conceptual terrain of modularity within
synthetic biology to that of *evolvability*. How can evolvability be characterized in
synthetic biology[7]? I suggest that it may be best understood as the capacity of a
population, organism, device, part, or pathway to change over time—that evolvability is the facilitated variation of self-organized systems (cf. Calcott 2014). Conceived
in this way it can serve as an umbrella term under which natural, artificial, and
synthetic change over time can be covered.

Evolvability relies on phenotypes being both plastic and stable. Phenotypes of
organisms are plastic insofar as they are responsive to the continual variation within
their environment. Phenotypes of organisms are stable insofar as they may develop

---

[7] For a recent discussion of evolvability and synthetic engineering that is complementary to the one
presented here, see Calcott (2014).

reliably despite changes in the resources available to the organisms. Novel phenotypes arise through the rearranging or recombining of ancestral phenotypes by the organism (West-Eberhard 2005: 6543). Organisms are able to coordinate the resources used in their own development because they are:

> richly endowed with a capacity for facilitating variation, a small input of random mutation would lead to a large output of viable phenotypic variation ... Instead of a brittle system, where every genetic change is either lethal or produces a rare improvement in fitness, we have a system where many genetic changes are tolerated with small phenotypic consequences and whereas others may have selective advantages, but are also tolerated because physiological adaptability suppresses lethality (Kirschner and Gerhart 2005: 226).

If facilitated variation is the mechanism of natural selection, then it would appear that the inputs are modules. Organismal modularity maintains robustness of pathways whilst allowing recombination of relatively autonomous entities. Facilitated variation can be understood as explaining the evolvability of organisms through its self-organizing processes and its relationships to genetic and extragenetic resources. In this sense, the evolvability of organisms may be underpinned by the specific modularity thesis that takes the nature of organisms to be organized modularly. Organisms may change their features depending on a number of factors. These changes can be attributable to the self-organizing capacities of organisms in selecting which resources, in what order, and in what combination, are used as causes of their own development. The organism could be understood to be the author of its own variability. Its generic capacities may constrain deleterious variation or enhance variability that may be advantageous to it. Using these capacities, organisms may either buffer or enhance any variations to their genetic or extragenetic resources or perturbations in their inter- or extra-cellular environmental resources (Kendig 2014a).

An organism has both constraints and de-constraints on the variability of its phenotype. Its capacity for variation allows it to maintain itself across a wide range of conditions. The organism's potential for different developmental variations lie in its capacity to self-organize its genetic and extragenetic resources. The organism's common stock of generic capacities and resources has been understood by Mary Jane West-Eberhard (2003: 146) as its "phenotypic repertoire". The organism's phenotypic repertoire includes a number of highly conserved core processes: intercellular signalling and cell sorting (Goodwin et al. 1993); the capacity for weak linkage, exploratory behaviour, spatial patterning, compartmentation and modularity of the body plan (cf. Kirschner and Gerhart 2005); the capacity of tissues to segment or form hollow tubes (cf. Minelli 2003); and the capacity to learn and remember (West-Eberhard 2003: Ch. 3, 7, 18). A thumbnail sketch of some of the ways in which variation is facilitated may be useful.

The capacity for weak linkage allows organisms to use a small number of mechanisms relying mostly on the ability to reconfigure these as and when necessary for different functions. But because these linkages are weak, they are often retraced and duplicated by other pathways or circuits to strengthen them. The use of weak linkages means that there is higher versatility which the organism can use to alter these pathways when necessary. They offer a socket and plug model that facilitates interchangeability of different modules. If these were strong, static and unchanging linkages, the organism would be arguably less fit to adjust to different environmental situations (Kirschner and Gerhart 2005: 136–7). In the course of evolution small changes result in wide variability and novelty. Weak linkages can be formed in many kinds of interactions, such as those between cells, cell populations, tissues, organs, organ systems, and behaviours. In this way, they confer a standardized way of connecting different modules to one another. Because these linkages are weak, individual organs, cells, tissues, or behaviours may also change independently. The exploratory behaviour of organisms is their responsiveness to different inputs and outputs. In building such things as neural networks or circuits, the organism begins by constructing a large number of alternative pathways. The best of these alternative pathways are selected and stabilized.

A conserved body plan – one that is retained over a succession of individual organisms – enables independent variation of some features without adversely affecting others. It does so by compartmentation of the body plan into semi-autonomous functional and structural parts. This modularity of the organism's parts increases its capacity for variation as changes in one subunit do not greatly affect others and thereby reduces the possibility of lethal variations. The more modular these subunits become, the greater the possibility of variation and specialization of these structural and functional units within the organism. The generic capacity to learn through exploratory behaviour provided certain motivating factors, such as the absence or presence of certain resources needed in the construction of a particular phenotypic trait or the performance of certain processes or behaviours (e.g. metabolism, reproduction, locomotion, speech), enables the organism to vary its phenotype over its lifetime in a range of different contexts. Organisms may learn about their resources and environment by quorum sensing, by chemical cues or by virtue of their sensory organs. They may manipulate the objects within their immediate habitat, investigate new resources or interact with new organisms (e.g. prey, potential mates, carers, symbionts) within this habitat, or search for a new one. Certain types of activities associated with access to food, protection from weather, increased sociality or reproductivity, or fitness may result in some benefit or detriment to the organism). Activities that effectively increase or substantially decrease resources are remembered and repeated or avoided. These generic capacities allow organisms to vary their own development and the phenotypes they construct depending on how resources are used (Kendig 2014a).

4.1 How Can Engineered Metabolic Pathways Be Maintained?

Building on the discussion of facilitated variation in the last section, I now use this to return to discuss it in light of particular examples of metabolic engineering in microbial organisms to explain how pathways can be maintained in synthetic biology. Metabolic engineering depends on the discovery and investigation of natural metabolic pathways and the genetic elements that control them, on using that information to transform suitable host organisms for the desired orthogonal metabolism, and on optimization of the metabolic output. The most high-profile product of metabolic engineering resulted in bacteria and yeast cells that produce the anti-malarial drug artemisinin, reducing its cost of production compared to purification from the sweet wormwood plant (Martin et al. 2003). Another is the recent attempts to engineer cells to make more of what they may make already, (e.g., single-celled algae producing oil, or producing a variation on what they would normally produce). This is exemplified in the current research trend in publically and privately funded projects is to investigate the potential use of various species of algae for biofuel production (Kendig 2014b). Algae produce lipids (oil) as a byproduct of the process of photosynthesis. The hope is that once the means of harnessing this store of energy is found, algal biofuels may provide an inexpensive alternative source of fuel that can be produced with little more than sunlight, carbon dioxide, and a small amount of water. While advances in synthetic biology research and the understanding of algal alternatives increase, the scaling up of these fuels requires significant further research resolving the problems of system optimization and photosynthetic efficiency as well as solving ways of producing these synthetic biofuels in quantities suitable for commercial use (Georgianna and Mayfield 2012). Various species of bacteria and yeast have been considered as particularly suitable for research into the production of synthetic biofuels (Dellomonaco et al. 2010). Cyanobacteria are another that initially appears promising. Cyanobacteria, like *Synechocystis* sp. PCC 6803, can provide a highly efficient organic system for producing biofuels as they can convert solar energy and carbon dioxide into biofuel molecules (Wang et al. 2013). Cyanobacteria are particularly good candidates because they possess naturally occurring biosynthetic pathways that produce alkanes (a key component of gasoline, diesel, and jet fuel). At present, research into the use of cyanobacteria for synthetic biofuel production is still in the very early stages and well behind that of algae research. However, research focused on reconfiguring these to create an organism that produces alkanes or alkenes at a rate that is double that of the wild type has been shown to be possible. *Synechocystis* mutants have been constructed that overexpress alkane biosynthetic genes (Kendig 2014b). If their photosynthetic pathways were re-engineered, cyanobacteria may be able to produce alkanes or alkenes at a highly efficient rate (Wang et al. 2013).

But knowing how to sustain the orthogonal pathway over generations requires resolution of two problems. I refer to the first of these as the *optimization problem*. The optimization problem involves figuring out how synthetic biologists can enhance orthogonal metabolic output when they cannot know all of the variables that affect it in a host organism and its environment. I call the second problem the *natural selection problem*. How can the reduction in fitness in a population of organisms engineered for an orthogonal metabolic pathway be prevented from causing evolution away from the desired metabolic output phenotype? The native metabolism of all organisms is responsive to a vast array of internal and external variables including genotype, metabolite availability, pH, osmotic pressure, and temperature (Eckdahl et al. 2015).[8] When synthetic biologists introduce orthogonal metabolism

---

[8] For an extended discussion of this example see Kendig and Eckdahl (2017) and Eckdahl et al. (2015).

into a host organism, they can optimize metabolic output by taking into account the variables they know, but must ignore all of the variables they don't know and some that they cannot know. Synthetic biology researchers also struggle with the difficulty of trying to make sure that the organisms they engineer with the ability to carry out orthogonal metabolism continue to faithfully replicate that genetic capacity. If the production of the product or the maintenance of the pathway is too onerous for the cell, the metabolic pathway will (after a number of generations) be disposed of. Those organisms without the onerously produced product or taxing pathway will be selected for in preference to those constructed to function by the synthetic biologist. To sum up in one sentence: synthetic biology's biggest obstacle is natural selection.

4.2   Is Semisynthetic Evolvability Still Evolvability?

In the previous section, I suggested that evolvability may be understood as the capacity of an organism, device, part, or pathway to change over time—that evolvability is the facilitated variation of self-organized systems. Conceived of in this way, evolvability was able to be used as an umbrella concept under which natural, artificial, and synthetic change over time can be covered. I now want to question the legitimacy of this as a category or kind. I suggest that synthetic biological research, as well as the orthogonal products and processes that it creates, sits in a liminal position between engineered technological kinds (cf. Schyfter 2012) and natural (or in the language of synthetic biology "native") biological kinds. If this liminality exists, it presents difficulties with regard to the notion of kindhood if kindhood is understood to be a property of the contents of the world as already there for us. Why? To explain, I'll use the butchery metaphor often used to explain the nature of natural kinds as being understood as picking out the natural categories in the world as *carving nature at its joints*. Synthetic biological research and the orthogonal products and processes that arise from it would suggest a notion of kindhood that does not rely on the pre-carved up contents of the world but instead a notion that involves the role of the carvers and—following the discussion of how assembly methods may affect modularity—the ascription of the joints.

As such, the goal of understanding evolvability is directed to the potential to modify and optimize the modules, networks and systems which direct it. This takes evolvability to be a capacity that can be defined, represented, measured, intervened upon and optimized. We might do better to call this "semisynthetic evolvability" as it differs from natural or artificial evolvability in that the synthetic biology practitioner actively intervenes in the process of facilitated variation and employs the use of a suite of constructed parts and employs populations of organisms to solve the biological problem of optimization.

But the question remains: are semisynthetic evolvability and the umbrella concept of evolvability as facilitated variation of self-organized systems legitimate and veridical categories of evolvability? And how are we to judge them? According to Morange, "Selection in nature and evolution directed by biochemists and molecular biologists increasingly differ as technological progress is made in synthetic biology" (Morange 2009a: 370). At first blush, this would suggest that semisynthetic evolvability may not count as evolution at all. Later on, however, Morange provides an alternative account. He argues against his initial suggestion that synthetically tweaked evolution does not count as evolution. He goes on to argue that current understanding of evolution is based on an unwarranted assumption of univocalism that could be challenged by findings in synthetic biology:

> some of the extraordinary scenarios of evolution, designed by synthetic biologists, must be carefully considered by evolutionary biologists. Maybe the models provided by evolutionary biologists are too restrictive, leaving too much place to uniformitarianism, the hypothesis that the mechanisms of evolution have always remained the same (Morange 2009a: 374).

Put stronger, perhaps evolutionary biologists have just been getting it wrong. Natural evolvability is actually more like semisynthetic evolvability. So research into the latter may yield understanding of the former. Judged according to Morange's second suggestion, the umbrella category of evolvability can also be viewed as veridical insofar as it captures the heterogeneity of synthetic, artificial, and natural evolutionary mechanisms as well as their concomitant epistemic categories.

5 The Social Aspect of Scientific Investigation

In the foregoing, I have attempted to show that attending to the practices within synthetic biology—in particular to the reengineering of metabolic pathways—reveals the generation of knowledge-making categories and the delineation of modules within the manipulation of biological parts, processes, and systems. I now turn to the social aspect of the practitioners' work. Although the specific focus on metabolic engineering is new, doing so relies heavily on Marjorie Grene's much earlier identification of the work of scientists as being not just the subject of epistemology but also of ethical engagement. Writing in 1966, Grene states that scientists' "work" is "an instance of the recognition of responsible persons, a performance of the same general kind as the recognition of patterns, individuals, or persons" (Grene 1966: 223). Grene refers to the social nature of science as work that takes place in "social enterprises" (Grene 1985). But Grene is not alone in pointing to the ineliminability of the social within scientific inquiry. It has been characterized more recently in terms of "systems of practice" (Chang 2012), referred to in virtue ethics as "agent-based" interactions (Swanton 2003), been described as "social cognition" (Longino 1990), and in some sense, it has been much earlier cashed out in terms of the concept of "conviviality" (Polanyi 1962). I build on this formindable work, focusing on the social aspect of scientific investigation and the notion of a system of practice in order to identify the work of practitioners in a reticulated set of knowledge-making activities.

Following Grene, I take the work of scientists to rely on protocols as sets of explicit and implicit rules of action. The action of individual agents within the social group of scientists is not the work of idealized decision-makers but seems to be biased on individuals whose beliefs are not entirely self-generated. That is, they do not form in a vacuum but from within a particular context and environment. As such, scientists working within social enterprises are agents that act from a particular embedded standpoint. Grene's view seems underpinned by an agent-based view that investigates a perspective where action is always from a standpoint, but that the standpoint is located in social networks of knowledge producing practices. It is that place from which and as such a precondition for personal action. In this way Grene appears to rely on Polanyi's notion of agency (see Polanyi 1962, and in particular, see Mullins 2009 for a discussion of Polanyi's notion of agency). For Polanyi, science is only considered possible within a society that acknowledges it by affirming the sorts of questions, types of reasoning, and value of their pursuit, advancement, and transmission of knowledge. Polanyi (1962: 203) suggests that scientific knowledge is only possible because the sharing of knowledge is itself valued in society. That it is valued, makes science possible. He suggests that reliance on this social endorsement is pervasive—it is prerequisite to all epistemic and ethical knowledge. Polanyi illustrates the pervasiveness:

> I cannot speak of a scientific fact, of a word, of a poem or a boxing champion; of last week's murder or the Queen of England; of money or music or the fashion in hats, of what is just or unjust, trivial, amusing, boring or scandalous, without implying a reference to a consensus by which these matters are acknowledged—or denied to be—what I declare them to be. I must continually endorse the existing consensus or dissent from it to some degree, and in either case I express what I believe the consensus ought to be in respect to whatever I speak of (Polanyi 1962: 209).

More recently, the pervasiveness of the social aspect of scientific enquiry as prerequisite for objective knowledge is later explored by Longino (1990). Longino's interest is in characterizing the nature of knowledge acquisition, but she cautions,

> Because we think the goal of the scientist's practice is knowledge, it is tempting to follow tradition and seek solutions in abstract or universal rules. Refocussing on science as practice makes possible the second shift, which involves regarding scientific method as something practiced not primarily by individuals but by social groups (Longino 1990: 66).

The social nature of scientific practice that Longino suggests here emphasizes the intimate connection between the character of inquiry as a social and is prerequisite for what she refers to as "social cognition"—the idea that scientific inquiry is not simply an individual pursuit but an epistemic activity relying on the intersubjectivity of critical dialogue within scientific study. In doing so, she follows Grene (1966, 1985) in fleshing out the nature of what it means to be a social enterprise and what I suggest Chang (2012) later identifies as a system of practice. According to Chang, the search for an agent-free or context-free set of categories is simply one that is ill-founded in philosophy. This is because knowledge is always something bound within what he refers to as a "system of practice":

> A system of practice is formed by a coherent set of epistemic activities performed with a view to achieve certain aims…[A]s with coherence of each activity, it is the overall aims of a system of practice that define what it means for the system to be coherent. The coherence

of a system goes beyond mere consistency between the propositions involved in its activities: rather, coherence consists in various activities coming together in an effective way toward the achievement of the aims of the system (Chang 2012: 16).

By focusing on systems of practice, Chang foregrounds the activities and goals of scientists within the pursuit of knowledge rather than with the product of their activity—the observations statements and their truth value. It is not the veridicality of statements on their own that tells us what constitutes success in science. In order to understand truth or anything like it, it is necessary to investigate how these statements function within the knowledge-making activities of scientists. The assessing of truth or truthlikeness of any statement is valuable but context dependent (Chang 2012: 17–18). Truth can only be assessed within the constraints of the goals and by the values set out by practitioners that are for instance, working towards solving a particular problem. That is, what counts as a solution depends on what is determined to be the problem.

What appears to connect Grene's notion of social enterprises, Longino's social cognition, and Chang's systems of practice is a sense of the pervasiveness of the social nature of scientific investigation—a pervasiveness highlighted by Polanyi's notion of conviviality, social endorsement, and dissent. I explore the pervasiveness of the social by looking once again to Polanyi's notions in order to characterize an account of ethics. According to Polanyi, it is not the agreement of the community of individuals that leads to knowledge but the possibility of disagreement (Polanyi 1962). Being in a community provides the environment that makes disagreement possible. To disagree is to see oneself as being against something and in the possibility of communicating that disagreement to another. Applying a Polanyi-inspired approach to normative ethics, I suggest that normative ethical evaluations are the result of loops of ethical reflectiveness of agents in a multi-agent network of practitioners who construct the grounds for ethical knowledge through their intersubjective judgements. That is, normative ethical evaluations rely on what people think, what they do, how they do it, and how they communicate it to others. These evaluations are dependent on the epistemic capabilities of individuals multiply instantiated in networks of practitioners. In synthetic biology, these networks include human agents, but also their physical manipulations (e.g. measuring, weighing, running gels), mathematical modelling, proxied or remote tool use, objects of study (e.g. chassis organisms, BioBricks), and the extended social communities that they work within (e.g. research networks that span multiple institutions, iGEM competitions, international research networks). They are spatially, socially, and temporally extended.

5.1   Extended Agency Ethics

In the remaining, I briefly consider how synthetic biology practice reveals valuational as well as the epistemic and ontological categories of research. I suggest that ethical categories, like the knowledge-making practices, are formed alongside these within socially extended research groups and sketch an approach to understanding the generation of ethical knowledge. I refer to this form of ethical knowledge as "extended agency ethics" to emphasize the intersubjective nature of this social approach. Extended agency ethics can be understood as taking a naturalistic agent-based approach to ethics. I clarify the descriptors "extended" and "agency" one at a time before bringing them together. I refer to this approach as "extended" because this view relies on the extended-mind thesis in cognitive science (Clark 1995; Clark and Chalmers 1998). According to the extended mind thesis, thinking is not exclusively something that happens in the brain, it is something that is spatiotemporally extended (Clark and Chalmers 1998; Clark 2010). Mind includes brain but also includes tools used to aid thinking (e.g. language, culture, pencils, calculators, search engines, mobile phone apps, and other people). Knowledge is not just epistemologically embedded according to this view, but it shapes and is reciprocally shaped by our experiences in the world:

> we use intelligence to structure our environment so that we can succeed with less intelligence…it is the human brain plus these chunks of external scaffolding that finally constitutes the smart, rational inference engine that we call mind (Clark 1998: 180).

I use this extended mind thesis to suggest how ethical decision making is ineliminably connected to the research practiced in groups—but these are groups of individuals capable of acting. I refer to "agency" in the description of the approach to indicate that the role of the individual agent. This agent based view of action implies that we, as agents, are the locus of our activities, e.g. measuring, mapping, or running gels, in order to effect changes beyond merely measuring, mapping, or running gels. Agents act for and towards a purpose beyond those movements. They are able to do so because they have certain causal knowledge of the result of the performance of these actions.[9] The agents' activities are not reducible to the events of measuring, mapping, or running gels because it is the person's running a gel and not the gel's—to assume that these are the same thing is to reduce the agency of the agent to an event. Keeping in mind the ubiquity of the social as expressed by Polanyi, Grene, Longino, and Chang, extended agency ethics does not assume that intentionality is the sole domain of the individual nor that the decisions of the individual are autonomous. Individual scientists' work is generated through a social process and forms a kind of interactive extended agency. That is, their research activities are extended over the members of the research group and coordinated in virtue of their shared intentionality in the form of extended cognition. These shared intentions provide the grounds for normativity in scientists' social research interactions. Robert Wilson has recently suggested a view of extended agency akin to the one I have developed here and elsewhere, (Kendig 2016a), but within a separate context.[10] He provides a

---

[9] I follow Hornsby's account of irreducible agent causation here, (see Hornsby 2004: 11–14 for further discussion of agency in philosophy of action and Lowe 2009: 196–201 for further distinction between agent causation and event causation).

[10] Wilson's (2018) discussion of normativity is given in the context of the eugenics movement and in particular, within a critical analysis of the cognitive processes that lead to the marking of certain human variation as deficient and other variation as preferred within scientific practice.

particularly helpful way of understanding how sociocognitive intentions can be construed as being the basis for normativity:

> Normativity exists when there is a distinction between a correct, proper, or appropriate way for a process, event, or outcome to turn out, and an incorrect, improper or inappropriate such way. Like extended cognition, normativity arises in and through both non-human and human cognition; it is not solely a feature of our own species' activity. [L]ike extended cognition, the most familiar and robust forms of normativity are those that are the product of distinctly human practices and institutions that presuppose a kind of shared intentionality…So we have a kind of externally mediated, cognitively driven normativity, and it constitutes an important feature of human social life (Wilson 2018: 130).

Agency is, therefore, not restricted to the individual or to a limited number of individuals. It can extend to the social research network one participates within with its requisite values and practices. Extended agency provides a conception of the social character of scientific inquiry that attempts to make sense of it as research that is extended across events, environments, objects, and agents. My suggestion is that responsibility for the form of ethical evaluations is therefore also distributed across the system of agents within the network of practitioners.

So, what would a normative ethical decision-making process look like using extended agency ethics? In pursuing research to find a suitable organism to be used as a chassis for biofuel production, a team of researchers would use knowledge acquired from the related areas of research outputs of projects focusing on cyanobacteria, algae, and metabolism. They may use this broad investigative approach to narrow down the range of possible candidates for a chassis organism. They may initially investigate the current algae research and the metabolic pathway of the highly familiar, well-researched green algae, *Chlamydomonas reinhardtii*. Knowledge of the successes and problems associated with the reengineering of algae as a source of biofuel may lead to the choice of a cyanobacteria instead. Once a chassis is selected, they may focus attention on the synthetic construction of pathways that overexpress alkane biosynthetic genes. Following this, they may begin the task of generating stocks of the newly reengineered form of *Synechocystic sp.* In order to plan the most efficient scale-up ventures, economists as well as microbiologists may be contracted. Once enough product is produced, they may outsource some beta tests to chemical engineers for kinematic viscosity analysis of the cyanobacteria-based biofuel product. They may request assistance in testing combustibility from colleagues specializing in physical chemistry. Limitations on what can be known and what can be done may come from the reciprocal knowledge exchanged through these interactions as well as with the cultures and communities potentially affected by the production of products. The normative ethical decisions and projectable outcomes are obtained through and by these interactions. Knowledge of the organism being used, the marketability of the product, the scale of production, and the social and environmental impacts are all linked to the particular organism used. That is, the ethical evaluations of biofuel production varies depending on the organism used, the scaling applied, the prospects for environmental controls, expected social effects, communication of these, and the impact on local and world economics.

Extended agency ethics describes the integrated approach to ethical decision-making as one that requires careful and critical consideration of integrated research and technological activities in order to reasonably predict possible outcomes. Ethical decisions concerning the potential use of Synechocystic- based biofuel is not something that comes as purely either backward looking assessment or forward speculation. The determination of what should be done is not something that can be judged solely on the basis of weighing up consequences of action, nor on the basis of rule- following—categorical or otherwise. Instead, ethical considerations—such as the permissibility of the development of some new technology (e.g. cyanobacte-rial- based biofuels)—are determined according to the current research, the experiential data amassed, the practitioners' knowledge-making activities, and the potential for scaling up production of the biofuel products by industry. This means that the locus of normative agency and intentionality is distributed across the activities of research groups, tools, and the development of products.

In this way, extended agency provides an externalist view of justification. Justification for beliefs come not from the decision-making-inside-the-head version of internalism (as some set of brain-bounded intuitionism or reflective perception). Instead, a form of active externalism that distributes cognition socially and agency across spatiotemporal research practices is suggested to explain the nature of scientific inquiry and the acquisition of knowledge. The ability to form research questions and pursue this kind of research depends on the aims of the system of practitioners within a particular research environment. As such, extended agency radically revises the conditions under which beliefs are considered to be justified. Justification can be on the basis of reflection on decision making as a form of mental cognition, but according to extended agency, what is taken to be mental cognition is not restricted to the brain but instead goes beyond the skull and can extend to tools, practices, processes, and other researchers and social research groups—social cognition. Both epistemic credit and ethical culpability are distributed notions that extend beyond the individual human agent. If social cognition, the social aspect of scientific practice, and the extended mind thesis are taken seriously, mental states (usually restricted to the brain by internalist evidentialists) are extended not only beyond the brain and skin of the individual but to include other spatiotemporally distinct biological, technological, and socially extended entities. With this extended cognition, a requisite extended agency and normative ethics based on this actively externalist theory of evidence follows. In some meaningful sense, extended agency can be understood to be a kind of role-based approach to ethics that demarcates categories of valuation analogous to those epistemic categories (outlined in the first half of this chapter). That is, to act in a role is to act according to a category of activity or to follow a model or prototypic way of acting. It may be profitably understood as being akin to an Aristotelian notion of virtue or a context- driven valuation. The intimacy of epistemic and ethical knowing is explicitly articulated by a number of virtue ethicists (see in particular Swanton's 2003: 249 "virtues as prototypes"). Instead of understanding morality as being based on a set of rules, this approach takes virtues to be frameworks. These frameworks are built from interactions and in-practice experience that both shapes and is shaped by future interactions in the world (Swanton 2003: 279). Christine Swanton's "virtues as prototypes" can be seen as a more restricted version of my extended agency ethics. Whereas Swanton limits knowledge of the world to agent-based interactions, I extend it further to include agent-object based interactions. That is, I include interactions between humans, but also between humans and their objects, technological artefacts, and tools of investigation.

I hope my extension will help to build a bridge between philosophy of science and sociology of science. The bridge can be understood in terms of a shared locus of research: the study of interactions and associations between social actors. The interactions between social actors—whether they be technological artefact, machine, or human—consist of a system of valuing within these systems of scientific practices. The normative status of the interactions, objects, and human actors comes from the valuing of practitioners, the scientific methods they choose, activities that they participate in, the tools and measuring devices that they select in their field of study to use, and the meaning attached to the information gained from those tools. These all constitute sources of meaning within the use and interaction with them in various ways. The aim of the extended agency approach outlined here has been to begin to explain agent-object based interactions within scientific research groups. In the next section, I suggest that these object-human interactions can be best understood through a combined sociology and philosophy of science.

5.2   Bridging Philosophy of Science and Sociology of Science

In his pivotal paper, "Mixing humans and nonhumans together: the sociology of a door-closer", Bruno Latour, (writing under the pseudonym "Jim Johnson"), suggests that studying the interaction between humans, machines, and tools should be the remit of a more widely extended approach to the study of sociology of science. He argues that if things we commonly refer to as a *tool* or a *machine* affect the way we interact in the world, then they might also be considered social actors. Considering them as such would blur the line that is often drawn between what is "purely technical" and what is "purely social" (Latour 1988: 198). He shows by means of a series of examples, (e.g. a hydraulic door closer and a red stop light), that some technological objects prescribe the behaviours of humans who interact with these artefacts. He focuses in particular on tools and technological objects that stand in for humans or take on the work of humans, (e.g. the hydraulic door opener that takes on the job of a human groom who opens the door for us, or a traffic light that takes on the job of a police officer who signals that we should stop in traffic). Although they are not human, the hydraulic door opener and the traffic light determine the norms of behaviour considered appropriate when we encounter them and sets the normative terms of interaction. That is, we interact with the non-human door-closer and the traffic light in terms of the roles each occupies. Knowing the role these technological objects play means that we also know what actions we should take in response to them. They bring with them the norms for how to interact with them. That these technological artefacts might be informed by our interaction with a polite human opening a door for us or a police officer who directs traffic in an area where no traffic lights exist does not seem to count against the normative interaction present when we drive up to a light and act according to certain norms. That the light is not a person doesn't mean we fail to understand that we should stop when it changes to red. The object in some sense normatively frames our experiences with it. It interacts with us in terms of how it interprets our role in the interaction. Latour calls the normativity that arises from the machine's presuppositions about the role of the human user, "prescriptivism".

> *Prescriptivism* is whatever a scene presupposes from its *transcribed*[11] actors and authors (this is very much like 'role expectation' in sociology, except that it may be inscribed or encoded in the machine). For instance, a Renaissance Italian painting is designed to be viewed from a specific angle of view prescribed by the vanishing lines, exactly like a traffic light expects that its users will watch it from the street and not sideways. In the same way as they presuppose a user, traffic lights presuppose that there is someone who has regulated the lights so that they have a regular rhythm (Latour 1988: 306).

Latour suggests we consider interactions with scientific objects and technological artefacts (and not just people) to be social interactions. As such, he argues that sociology should treat the material or technological objects we relate to as not just of peripheral social interest but of direct sociological significance. Doing so requires what Andrew Pickering calls "a decentering of the social relative to the material and the conceptual, in terms of both objects of analysis and explanatory formats" (Pickering 2005: 352). The current chapter attempts to follow Pickering's approach to decentered sociology by "exemplifying empirically and theoretically what a decentered sociology can look like, in the hopes of encouraging others to follow" (Pickering 2005: 353). In the above, I have attempted to provide a case study of how the tools and assembly methods chosen by synthetic biology practitioners may be themselves a sociological topic as well as a philosophical topic. They affect the process of knowledge production as well as the normative judgements of practitioners about the products produced within the collaborative social practice of research itself. Sociological investigation of the material and conceptual tools that affect knowledge production within labs may sheds light on how tools and decisions about tool use within systems of practitioners are made. In this way, normative as well as epistemic aspects of scientific research seem to require an understanding of social relations and not just a philosophical explanation of knowledge production. A decentered social theory and a decentered philosophical theory that focus on the role of technological objects may provide the best account of the normative interactions within collaborative groups of scientists.

---

[11] Latour uses "transcription" and "inscription" to explain the transition from a less durable delegated agent to perform an action to a more reliable one. For instance, "the replacement of a policeman by a traffic-light" is an instance where the traffic light is delegated the work that was done by a police officer. In Latour (1988), the focus is on descriptions of meaning that these actors play within a particular semiotic script. How actors are defined and what is meant by their roles in a particular scenario.

6 Concluding Remarks

The account of both epistemological and ethical ineliminability of scientific practice developed here aims to provide non-reductive grounding to knowledge in social cognition and within social enterprises. The possibility of critical dissent by the individual practitioner against the protocols, types of reasoning, modes of knowledge transmission, or generation of research questions affirmed by the community is what is required for knowledge. That requires the role of the practitioner as individual agent to be something that is irreducible to the social group or system of practice itself. Rather than a relationship of reduction of the individual to the social, the individual practitioner as agent can been understood to be working towards the goal of scientific knowledge and the performance of it using diverse methods. The practitioner extends her knowledge by making connections with other practitioners, making use of protocols endorsed by the community, using reliable model organisms to make her hypotheses, critically responding to accepted routes of knowledge generation, and networking her resources together in the performance of scientific inquiries. As an agent, she brings these resources together and in turn becomes a resource within the system of practice that she works within.

Investigating a practice-based route of acquiring knowledge within synthetic biology presents an alternative way of exploring traditional metaphysical questions in philosophy such as: What are the kinds of things synthetic biology produces? What is the relationship between parts and wholes? It also introduces new questions about what it is to be a part or indeed the property of partness itself within a new field; questions not just of interest to metaphysicians of science. Instead of seeking to understand the use of epistemological and ontological categories in practice from the premise that their existence can be known a priori or contained within the theoretical framework of the discipline that uses them, this approach runs in the opposite direction. It instead suggests that categories come into being in practice and from these categories-in-use theoretical concepts, notions of causal directionality, functional architectures, and normative valuation arise through the active engagement of systems of practitioners.

Acknowledgements Research for this chapter was partially funded by the National Science Foundation Division of Molecular and Cellular Biosciences (MCB), BIOMAPS: Modular Programmed Evolution of Bacteria for Optimization of Metabolic Pathways, Grant No. MCB-1329350, Amendment No. 001, Proposal No. MCB-1417799. Thanks to Todd Eckdahl, Jeff Poet, Malcolm Campbell, and Laurie Heyer for sharing their insights and expertise in synthetic biology with me during research for the project. Special thanks go to Phil Mullins for many lively discussions about Polanyi and for encouragement in the early stages of writing this chapter. I am also very grateful to Hauke Riesch, Brian Rappert, and Thomas Reydon for their feedback on earlier versions of the manuscript.

References

Brent, R. 2004. A Partnership Between Biology and Engineering. *Nature Biotechnology* 22: 1211–1214.

Calcott, B. 2014. Engineering and Evolvability. *Biology and Philosophy* 29 (3): 293–313.

Chang, H. 2004. *Inventing Temperature: Measurement and Scientific Progress*. New York: Oxford University Press.

———. 2011. How Historical Experiments Can Improve Scientific Knowledge and Science Education: The Cases of Boiling Water and Electrochemistry. *Science & Education* 20: 317–341.

———. 2012. *Is water H2O?* New York: Springer.

———. 2014. Units of Analysis in Philosophy of Science After the Practice Turn. In *Science After the Practice Turn in the Philosophy, History, and Social Studies of Science*, ed. L. Soler, S. Zwart, M. Lynch, and V. Israel-Jost. London: Routledge.

Chang, H. 2016. The rising of chemical natural kinds through epistemic iteration. In *Natural kinds and classification in scientific practice*, ed. C. Kendig, 33–46. Abingdon/ New York: Routledge.

Clark, A. 1995. I Am John's Brain. *Journal of Consciousness Studies* 2 (2): 144–148.

———. 2010. *Supersizing the Mind: Embodiment, Action, and Cognitive Extension*. Oxford: Oxford University Press.

———. 1998. *Being there: Putting brain, body, and world together again*. Cambridge: MIT Press.

Clark, A., and D. Chalmers. 1998. The Extended Mind. *Analysis* 58: 7–19.

Dellomonaco, D., F. Fava, and R. Gonzolez. 2010. The Path to Next Generation Biofuels: Successes and Challenges in the Era of Synthetic Biology. *Microbial Cell Factories* 9: 3. https://doi.org/10.1186/1475-2859-9-3.

De Regt, H., S. Leonelli, and K. Eigner, eds. 2009. *Scientific Understanding: A Philosophical Perspective*. Pittsburgh: University of Pittsburgh Press.

Dupré, J. 1993. *The Disorder of Things: Metaphysical Foundations of the Disunity of Science*. Cambridge, MA: Harvard University Press.

———. 2006. *Humans and Other Animals*. Oxford: Clarendon Press.

Eckdahl, T.T., A.M. Campbell, L.J. Heyer, J.L. Poet, D.N. Blauch, N.L. Snyder, et al. 2015. Programmed Evolution for Optimization of Orthogonal Metabolic Output in Bacteria. *PLoS One* 10 (2): e0118322. https://doi.org/10.1371/journal.pone.0118322.

Endy, D. 2005. Foundations for Engineering Biology. *Nature* 438 (24): 449–453.

Erwin, D., and E. Davidson. 2009. The Evolution of Hierarchical Gene Regulatory Networks. *Nature Reviews Genetics* 10: 141–148.

Georgianna, R., and S. Mayfield. 2012. Exploiting Diversity and Synthetic Biology for the Production of Algal Biofuels. *Nature* 488: 329–335.

Gibson, D., J. Glass, C. Lartigue, V. Noskov, R.-Y. Chuang, M. Algire, et al. 2010. Creation of a Bacterial Cell Controlled by a Chemically Synthesized Genome. *Science* 329 (5987): 52–56.

Goodwin, B., S. Kauffman, and J. Murray. 1993. Is Morphogenesis an Intrinsically Robust Process? *Journal of Theoretical Biology* 163: 35–144.

Grene, M. 1966. *The Knower and the Known*. Berkeley: University of California Press.

———. 1985. Perception, Interpretation, and the Sciences: Toward a New Philosophy of Science. In *Evolution at a Crossroads*, ed. D.J. Depew and B.H. Weber, 1–20. Cambridge, MA: MIT Press.

Hacking, I. 1992. The Self-Vindication of the Laboratory Sciences. In *Science as Practice and Culture*, ed. A. Pickering, 29–64. Chicago: University of Chicago Press.

———. 1995. The Looping Effects of Human Kinds. In *Causal Cognition: A Multidisciplinary Debate*, ed. D. Sperber, D. Premack, and A.J. Premack, 351–394. New York: Clarendon Press.

Hornsby, J. 2004. Agency and Actions. In *Agency and Action*, ed. J. Hyman and H. Steward, 1–23. Cambridge: Cambridge University Press.

Hylton, W. 2012. Craig Venter's bugs might save the world. *The New York Times*, March 6.

Keller, E.F. 2009. Knowledge as Making, Making as Knowing: The Many Lives of Synthetic Biology. *Biological Theory* 4 (4): 333–339.

Kendig, C. 2014a. Towards a Multidimensional Metaconception of Species. *Ratio* 27 (2): 155–172.

———. 2014b. Synthetic Biology and Biofuels. In *Encyclopedia of Food and Agricultural Ethics*, ed. P.B. Thompson and D.M. Kaplan, 1695–1703. New York: Springer.

———. 2016a. What Is Proof of Concept Research and How Does It Generate Epistemic and Ethical Categories for Future Scientific Practice? *Science and Engineering Ethics* 22 (3): 735–753. https://doi.org/10.1007/s11948-015-9654-0.

———. 2016b. Activities of kinding in scientific practice. In *Natural kinds and classification in scientific practice*, ed. C. Kendig, 1–13. Abingdon/New York: Routledge.

———., ed. 2016c. *Natural Kinds and Classification in Scientific Practice*. Abingdon/New York: Routledge.

Kendig, C., and T.T. Eckdahl. 2017. Reengineering metaphysics: Modularity, parthood, and evolvability in metabolic engineering. Special issue: Ontologies of living beings (eds A.M. Ferner and Thomas Pradeu) *Philosophy, Theory, and Practice in Biology* 9(8). https://doi.org/10.3998/ptb.6959004.0009.008.

Kirschner, M., and J. Gerhart. 2005. *The Plausibility of Life*. New Haven: Yale University Press.

Knight, T. 2003. *Idempotent Vector Design for Standard Assembly of Biobricks*. MIT Synthetic Biology Working Group.

Latour, B./Johnson, J. 1988. Mixing Humans and Nonhumans Together: The Sociology of a Door-Closer. *Social Problems* 35 (3): 298–310.

Longino, H. 1990. *Science as Social Knowledge: Values and Objectivity in Scientific Inquiry*. Princeton: Princeton University Press.

Lowe, E.J. 2009. *A Survey of Metaphysics*. Oxford: Oxford University Press.

Martin, V.J., D.J. Pitera, S.T. Withers, J.D. Newman, and J.D. Keasling. 2003. Engineering a Mevalonate Pathway in Escherichia coli for Production of Terpenoids. *Nature Biotechnology* 21 (7): 796–802.

Minelli, A. 2003. *The development of animal form*. Cambridge: Cambridge University Press.

Morange, M. 2009a. Synthetic Biology: A Bridge Between Functional and Evolutionary Biology. *Biological Theory* 4 (4): 368–377.

———. 2009b. A Critical Perspective on Synthetic Biology. *HYLE* 15 (1): 21–30.

Mullins, P. 2009. Polanyi on Agency and Some Links to MacMurray. *Appraisal* 7 (3): 11.

O'Malley, M. 2009. Making Knowledge in Synthetic Biology: Design Meets Kludge. *Biological Theory* 4 (4): 378–389.

O'Malley, M., A. Powell, J. Davies, and J. Calvert. 2008. Knowledge-Making Distinctions in Synthetic Biology. *BioEssays* 30: 57–65.

Pickering, A. 2005. Decentering Sociology: Synthetic Dyes and Social Theory. *Perspectives on Science* 13 (3): 352–405.

Polanyi, M. 1962. *Personal Knowledge: Towards a Post-critical Philosophy*. Chicago: University of Chicago Press.

Rheinberger, H.-J. 2005. A Reply to David Bloor: Toward a Sociology of Epistemic Things. *Perspectives on Science* 13: 406–410.

Rouse, J. 1996. *Engaging Science: How to Understand Its Practices Philosophically*. Ithaca/London: Cornell University Press.

———. 2003. *How Scientific Practices Matter*. Chicago: University of Chicago Press.

Schlosser, G., and G. Wagner, eds. 2004. *Modularity in Development and Evolution*. Chicago: University of Chicago Press.

Schyfter, P. 2012. Technological Biology? Things and Kinds in Synthetic Biology. *Biology and Philosophy* 27: 29–48.

Soler, L., ed. 2012. *Characterizing the Robustness of Science: After the Practice Turn in Philosophy of Science*. Volume 292 Boston Studies in the Philosophy of Science. New York: Springer.

Soler, L., S. Zwart, M. Lynch, and V. Israel-Jost, eds. 2014. *Science After the Practice Turn in the Philosophy, History, and Social Studies of Science*. London: Routledge.

Sprinzak, D., and M. Elowitz. 2005. Reconstruction of Genetic Circuits. *Nature* 438 (7067): 443–448.

Swanton, C. 2003. *Virtue Ethics: A Pluralistic View*. Oxford: Oxford University Press.

Wagner, G., M. Pavlicev, and J. Cheverud. 2007. The Road to Modularity. *Nature Reviews Genetics* 8: 921–931.

Wang, W., X. Liu, and X. Lu. 2013. Engineering Cyanobacteria to Improve Photosynthetic Production of Alka(e)nes. *Biotechnology for Biofuels* 6: 69. http://www.biotechnologyforbiofuels.com/content/6/1/69. Accessed 1 Nov 2015.

West-Eberhard, M. 2003. *Developmental Plasticity and Evolution*. Oxford: Oxford University Press.

———. 2005. Developmental plasticity and the origin of species differences. *PNAS* 102: 6543–6549.

Wilson, R. 2018. *The Eugenic Mind Project*, 99–140. Cambridge/London: MIT Press.

Ethics and Citizen Participation in the uBiome Institutional Review Board Debate: Some Reflections on Social and Normative Analyses

Lorenzo Del Savio

1 Citizen Science and Crowdsourced Biomedicine

Digital technologies have drastically reduced the cost of recruitment of non-professional scientists for a variety of research tasks such as data and sample collection, data coding and problem solving. Open calls for participation in research can be advertised online and participants can be aggregated through web platforms. As a result, participation in scientific research is becoming increasingly open to citizens. This is sometimes called "citizen science". Key epistemic benefits of citizen science include the access of researchers to a large quantity of labour for the production and analysis of massive quantities of data, an efficient search for appropriate and rare skills that match particular research problems, and an increase in knowledge and experience diversity available for research (Sauermann and Franzoni 2015). Further benefits may also include social goods such as increased scientific literacy and expanded public control on science and research agenda setting (Del Savio et al. 2016).

Citizen science has firstly spread in disciplines as astronomy, ornithology and ecology, where a large number of amateurs were ready to join the activities of researchers. Several research projects in the life sciences are also starting to devolve substantial parts of their activities to the 'crowd' through online platforms. Crowdsourcing and 'citizen science' in these disciplines have attracted considerable attention because they involve human subjects research conducted in hitherto unexplored settings, which may involve their own specific ethical issues (O'Connor 2013; Vayena et al. 2015). In biomedicine, there is by now a wide and diverse array of such participatory projects. In the case of the health sciences, such initiatives are also called "patient-centric initiatives" although when patients and healthy

L. Del Savio (✉)
Christian-Albrechts-Universität zu Kiel, Kiel, Germany

© Springer International Publishing AG, part of Springer Nature 2018
H. Riesch et al. (eds.), *Philosophies and Sociologies of Bioethics*,
https://doi.org/10.1007/978-3-319-92738-1_4

individuals are involved in large-scale projects only as funders or donors of samples, data or information such label is questionable.

Examples of crowdsourced research in the health sciences include large scale genomics project recruiting on-line such as the *Genographic project*, open source genomics studies conducted on platforms such as *Genomera* and patient social networks facilitating the formation of patients´ communities that can initiate their own research. In the *American Gut Project* and its transatlantic branch, the *British Gut Project*, researchers collect bacteria samples from volunteers recruited online, who also provide phenotypic and lifestyle-related information, and funding. Other examples of crowdsourced initiatives are social networks such as *PatientsLikeMe* or *Curetogether*, which collect data about patient and their disease histories. These are also platforms that facilitate systematic self-experimentation with drugs. In addition, internet users play online games that help biomedical scientists with complex analytical tasks, e.g. *Eyewire* or *FoldIt*.

Most of these initiatives are conducted in fairly traditional institutional settings: they are initiated and overseen by professional scientists who outsource to the crowd those parts of the project that, despite being modular, cannot be automatized and requires substantial humanpower, wide geographical distribution or both. Only in few cases these research projects are either initiated by non-professional clinician/scientists or conducted by firms that have only recently started participating in biomedical research. These startups are able to enter the marketplace of ideas only through the mobilization of non-professional volunteers. These are arguably the cases that pose the most interesting ethical and social questions and have attracted attention from both Bioethics and Social Studies of Science. In the following, I reconstruct one of these debates on citizen science, pertaining to the microbiome sequencing service *uBiome* (Sect. 2). I will then situate such debate in its socio-economic context (Sect. 3). Finally, I will reflect on what a dialogue between Bioethics and Science and Technology Studies can deliver (Sect. 4). The analyses and indeed even the policy recommendations provided by scholars belonging to these two disciplines are often distinctly different, and I will use the current case study to describe their complementarities and tensions between.

2   The Debate on *uBiome* Ethical Review

Human microbiomics is the study of micro-organisms that live as commensals, symbionts or parasites of human beings and are thought to play important roles in development, health and disease. *uBiome* is an US-based startup that offers a gut bacteria sequencing service to consumers to entice data donation. It aims to establish a genomic repository for microbiomics. *uBiome* can be thought as a private and for profit version of the *American and British Gut Project* insofar customers can opt for one of the two platforms to obtain very similar data regarding their microbiome.

What happens to customers who subscribe to one of the two services? Upon payment (which covers the full cost of shipping, sequencing and analysis), they receive a swabbing kit that can be used to collect and ship bacteria samples, usually but not only from stool. Samples are then analysed and data given back to customers, with an overview of their microbiome composition. Microbiome data have no clinical utility at the moment, and researchers duly disclose this important limitation to users, whom however can compare their results with those of others and even with those of celebrities that have published their own results online.

Both projects run the direct-to-participants sequencing service as a strategy to obtain data that increase the volume of their own datasets. The objectives and rationales of the two projects are however different. The *American Gut Project* is committed to release data to researchers through the *earthmicrobiome* database project. *uBiome*'s value is based instead on the prospective sale of datasets to pharmacological companies and other biomedical firms, and as such it must ensure scarcity of the data it generates. Both operations claim to advance microbiomic research, but they function under very different assumptions. The *American Gut Project* embodies an open access approach according to which data should be shared to scientists as much as quality and ethical considerations make it possible in order to accelerate the production of knowledge. *uBiome* must instead retain control on datasets to ensure profitability and re-invest profits into their own activities, including valuable further research (Tempini and Del Savio 2018). They both claim to contribute to the science of microbiomics, although they opted for two different business model and data property regimes.

The scientific aim of *uBiome* is advertised in the website and arguably plays an important role in the recruitment of participants. These aims have attracted considerable attention: if presented as citizen science project, a sequencing service becomes as such research on human subjects, which poses particular ethical issues (Emanuel et al. 2000). In particular, human subject research, when conducted by a publicly funded institution or with publicly funded grants, is legally mandate in several countries to undergo official ethical reviews, which in the US and other countries take the name of *Institutional Review Board* (IRB) approval. These are committees representing scientific, legal and ethical expertise that evaluate whether a research projects meet current ethical standards. Major research centres run their own IRBs, but agencies providing ethical approval for third parties for fees do exists. In 2013, a debate has taken place on whether *uBiome* needed to undergo ethical review.

In the Frequently Asked Questions page, *uBiome* website formerly (November 2015) declared that the company was "working with an independent IRB to provide ethical oversight" and that they "will provide complete details when this process is complete". *uBiome* promised that "our IRB will be completed before any kits are sent out and before any consent forms are signed", alluding to the fact that the approval process was still in the making.[1] Moreover, *uBiome* claimed that "IRB

---

[1] The process has been completed as the website now (September 2015) states as follows: "An Institutional Review Board (IRB) is a committee that approves, oversees and reviews biomedical

approval is not required for us to provide our primary service of microbiome compositional analysis and interpretation for private parties" and that "our sample collection is part of a service and our research study is a meta-analysis of de-identified data, which is technically exempt from IRB". They also claimed that no IRB is legally mandated for their operations as no public money is involved. These declarations do not appear any longer in the website (September 2016).

These claims generated a host of reactions on the web and especially in some science blogs. This is interesting in its own respect as it testifies the capacity of reaction of a system of ethical vigilance based on online blogs. These reactions may have inspired some of the reflections on forms of "IRB 2.0" for citizen science that the debate later yielded (Vayena and Tasioulas 2013a). In particular, a user whose nickname is "physioprof" directly attacked the reasoning of *uBiome* so: "Their own Web site and their third-party "crowd-funding" site make it very clear that one of the main purposes–perhaps the primary purpose, on a reasonable reading–of their project and one of the major benefits to the participants they have already been recruiting is the aggregation and statistical analysis of the individual results so as to contribute novel information to the scientific understanding of the relationship between the human microbiome and human health/disease." (Physioprof 2013). Physioprof is pointing out that under any reasonable reading, *uBiome* claims that the intended use data includes research purposes, over and above the sequencing services offered to customers. Furthermore, the blogger explains that "meta-analysis exemption [from IRB] applies solely to de-identified data obtained from other already-performed human subjects studies *that themselves have been performed subject to appropriate IRB oversight*". In particular, "this exemption cannot possibly be applicable to a situation where the directors of a study collect human tissue/fluid samples, perform a biological analysis, and claim its "primary purpose" is a service to the participants, and that their subsequent aggregation and statistical analysis of the data that they collected is nothing but a meta-analysis", which is what *uBiome* declares in the website. According to the blogger, there was both a legal issue with the operations of *uBiome* (they must have sought IRB before initiating the sample collection), and a moral issue, as *uBiome* may have even unwittingly exposed participants to risks, such as for instance unwanted data disclosure.

Jessica Richman, the co-CEO of the company, replied quite accommodatingly in a comment to the blogpost that: "Although IRB approval is not required for us to provide our primary service of microbiome compositional analysis and interpretation for private parties, we have planned to do so once our crowdfunding campaign ended, which occurred two days ago. I appreciate the dialog here on this blog. Although I think the tone of the original blog post is unnecessarily combative, we are glad for the focus on ethical oversight, and share the goal that citizen science be conducted in a thoughtful, privacy-aware, and ethical manner." (Physioprof 2013).

---

and behavioral research involving humans. They determine whether or not research should be done, and they advocate for the individual participant (like you!) uBiome has received research study approval from E&I Review Services."

The debate may have ended there, but instead it quickly escalated. Few months later the debate continued in other blogs, appeared in the pages of Scientific American and *The Wall Street Journal* and eventually in *Nature Biotechnology*. Physioprof, who arguably initiated the debate, reminded to *uBiome* staff of the rationale behind the IRB governance system. There are reasons to be skeptical towards pure good intentions-based strategies to ensure that research on human subjects is ethical. Researchers have their own biases and prejudices and may over-look unethical aspects of their researches and in the past gross abuses have been committed in the name of the advancement of science. Bioethicists also intervened in the debate. For example, the Journal of Medical Ethics reminded the discussants of the principles of IRB approval contained in the World Medical Association's declaration of Helsinki: "The research protocol must be submitted for consideration, comment, guidance and approval to a research ethics committee before the study begins. This committee must be independent of the researcher, the sponsor and any other undue influence" (Graber and Graber 2013).

The *Wall Street Journal*, in an article revealingly entitled "Ethicists are pushing back against 'citizen scientists' who want to do medical research on themselves", offered a number of counterarguments (Dockser Marcus 2014). The key idea behind the article is further explored in a counter blog post by Richman and the other co-founder of *uBiome* Zachary Apte in the *Scientific American*. They complain that legal requirements for ethical approval of research on human subjects (IRB) are biased toward established research institutions that can afford to pay for IRBs, thereby "stifling innovation". Legal restrictions pertaining to research ethics constitute in their view a barrier to market access for startups, a barrier that generates innovation-unfriendly monopolies. In addition, they accused critics of citizen science of being paternalistic, as arguably participatory projects need not be subjected to the same ethical constraints of traditional projects since subjects are themselves in charge and able to consent or object to the various aspects of the research.

The debate was eventually taken up by bioethics scholars in *Nature Biotechnology*. Vayena and Tasioulas argued that patient-led research "is on the horns of a dilemma. On the one hand, given the risks potentially involved in research with human subjects, PLR's future sustainability depends upon instituting effective mechanisms of ethical oversight that are capable of securing the trust of participants and other stakeholders. On the other hand, the wholesale imposition of the standard ethics review that is legally required in the case of ILR—a procedure involving scrutiny by an institutional review board (IRB) and other forms of ethical oversight—threatens to stifle PLR, subjecting it to a regulatory straitjacket that may act as a disincentive to adoption and innovation" (Pg. 786). Both scholars considered the *uBiome* project "a vivid illustration of the complexity" of the dilemma they had pointed at earlier.

There are two problems with their opinion however, and they both may have been avoided with the help of a gaze on the debate inspired by science and technology studies and their attention to the social embedment of new technologies. Firstly, the analysts decidedly include *uBiome* among "participant-led" projects even though the label is in this case questionable, as no amount of control on research planning is devolved to participants. Secondly, they take up the frame of the debate as it is framed by one party, namely *uBiome* entrepreneurs opposing innovation to the "regulatory straitjacket". The opinion has also some internal tensions: regulation cannot at the same time promote the flourishing of participant-led research by sustaining trust among stakeholders and hinder innovation by disincentivizing creative developments, at least if we assume that trust among stakeholders is itself a key ingredient of the processes of innovation.

3 Crowdsourcing Research in the Contemporary Bioeconomy

Paying attention to the broader societal context allows the analyst to keep due distance from the rhetoric and views of the stakeholders. In a TED talk Richman describes the scenario of patient-led research as one where "technological forces are bringing us together to do science" (Richman 2013). She predicts that scientists will become facilitator of this form of research "setting up structures to integrate citizens into science". Her vision includes the employment of the enormous research potentials of non-professionals who have so far not participated in science: people in the developing countries, people who are not of "the right gender and the right skin color", or all those "not fortunate enough" to take part in research. Science can be made more innovative if it becomes a "democratized open system where anyone can participate" as a "human beings with the capacity to understand the world". In a *crescendo*, she imagines a world where everybody is able "to set the research agenda" and to pursue interesting lines of inquiry, a crescendo that is arguably at odds with the forms of participation that *uBiome* envisages for participants: sending stool, oral and/or genital swabs (including samples from pets) along with personal data about lifestyle and diseases to a sequencing centre. This is in one sense a way of "turning anecdotes in data", as Richman puts it, but non-professional certainly will not be in charge of research projects, nor they will be able to set research agendas. This observation alone refutes the claim that is made on behalf of "participatory" projects that ethical standards can be loosened as research subjects themselves decides on key aspects of these projects.

Science and technology scholars have been always keenly aware of the importance of citizen participation in research, and indeed the advocacy of public participation in research has functioned as the implicit political theory of many STS theorists (Thorpe 2010). Life sciences have also occupied a place of particular importance in these discussions, as patient movements have pioneered the campaign to reclaim the importance of "lay" people contributions to research and other activities hitherto reserved to elites (Madeleine and Rabeharisoa 2012). Citizen science and patient-led research has indeed become widespread in the life sciences by merging the increasing empowerment of patients in the management of their conditions (Vayena and Tasioulas 2013b) and the drive towards a precise, predictive, personalized, preventive and participatory medicine (Hood and Friend 2011; Prainsack 2014). However, while the aforementioned works on citizen participation have stressed the disruptive potential of lay people subverting hierarchies of knowledge production, there is now a renewed awareness of the possible limitations of such values and ideals, especially as they are seen as being increasingly swallowed up by corporate and managerial discourses. Sociologist Thorpe has argued that the success of Science and Technology Studies advocacy of participation among policy makers (a success whose nature has however been contested, Wynne 2006) has been due to its suitability to a regime of production that is based upon the productive capacities of the crowds. Thorpe and colleagues rely on analyses of post-industrial societies based on the diffusion of "affective work", and on the increasing importance of individualized inputs vis-à-vis standardized production (Thorpe and Gregory 2010). Taking up such analyses, Melinda Cooper has argued that citizen participation in some leading citizen science projects are paradigmatic cases where certain powerful corporate groups extract value from the contribution of unwitting citizens (Cooper 2008). In addition to these critical uptakes, other scholars have acknowledged the utility that such projects may generate for participants (e.g. in terms of self-knowledge, fun, satisfaction), but still inscribe their strategies in an economic context based on big-data and human personal information where it is vital to mobilize participants to obtain data and samples. In particular, they have noted that there has been a shift in rhetoric around participation. Formerly, firms tended to mobilize participants on the basis of an imaginary of an entrepreneurial self who takes care of her own health and optimize her own lifestyle (e.g. direct-to-consumer genetic testing services as 23&me). Now it is more common to appeal to an imaginary of altruistic self that contributes to research as she participates elsewhere in the so-called sharing economy (Tutton and Prainsack 2011).

More recently, we argued that recruiting participants online in projects as *uBiome* can be seen as illustrating a "Schumpeterian strategy" (Del Savio et al. 2016), after Schumpeter's view of economic activity as incessant revolution promoted by innovative newcomers that disrupt existing firms and productive structures when they are not stifled by barriers that generate monopolies. *uBiome* is in fact able to collect biological samples although it lacks academic credentials and hence cannot rely on institutional channels of advertising and sampling (e.g. hospitals and other healthcare institutions) precisely on the bases of its rhetoric of citizen science and participation. For these newcomers, recruiting participants on the internet is a source of biological samples and data as much as ventures and other financial instruments dedicated to startups are a source of funding. Startups are normally not able to raise capital in ordinary capital markets, but they can employ, especially in the US, a range of financial instruments specifically designed for risky investments. These financial channels lower entrance barriers to markets and as such they are seen by some theorists as fostering innovation by increasing competition, and hence incentives for research and development among existing firms. In other words, crowdsourcing and crowdfunding can be conceived as a remedy to an inefficiency of the market due to barriers generated by lack of trust in newcomers. Again, the participatory nature of firms as *uBiome* should be at least qualified, as it does not necessarily correspond to participants´ control or ownership on research processes, data, and outcomes.

By paying attention to the social and economic context, these analyses are not meant to discredit the potential of citizen science to deliver important social goods. They do put however into question the use of words as "patient-led" (or "citizen-led") research. They also put into perspective the framing of the debate in terms of innovators versus regulators. Appeals to participation are often ambiguous and pose significant ethical challenges. A recent opinion of the group of *Ethics in science and new technologies* of the European Commission (European Group on Ethics in Science and New Technologies 2015) warns that "participation may elicit expectations as to greater transparency or accountability, but cannot necessarily provide it. The term "participatory" may be attributed to services where consent is ambiguous. It can be based on the extraction and sale of personal data, and where it concerns the extraction of profit or labour, act as a veiled form of exploitation." A key test of whether or not appeals to participation are being misused is based on the degree of control on data and research proceedings that is devolved to participants, which in the case of *uBiome* is arguably none. Knowledge of the background analyses on participatory science by science and technology scholars is in this case instrumental for robust ethical analysis.

4   Science and Technology Studies as Ethical Analysis

The underlying moral issues in the debate regarding the IRB of *uBiome* pertains to whether or not it is an ethical requirement to have crowdsourced research project overseen by Institutional Review Boards as it happens in the case of traditional human subject research projects. Traditional research ethics focuses on risks to participants, consent, disclosure of information, respect of research subjects, scientific value of research. Science and Technology Studies (STS) identify of a broader set of social goods that might be affected by such decision, in this case the nature of participatory projects in biomedicine, their place in contemporary life sciences and their innovation potential. This is a valuable contribution to ethical analysis: goods such as participation and the transformation of the research system are not easily visible from within the frameworks of research ethics, which most often involves a balancing between individual-level goods and harms against the public value of medical knowledge. A second important contribution of Science and Technology Studies is critically reflecting on what stakeholders themselves have to say on the ethical issues under discussion, in this case the coupled narratives of powerless innovators fettered by regulators and self-organizing citizens who participate in research and own substantial stakes in the process.

How should we describe the relationship between sociological and moral analysis on the base of this case study? It may be tempting to describe the respective roles of STS scholars and bioethics analysts in term of facts and values expertise. For instance, STS scholars would have a firm empirical grasps on – say – whether or not *uBiome* gives power to participants, and ethicists would then reflect on the consequences of this fact pertaining to what makes these researches ethical. Even if we let aside the complicated debates regarding whether the distinction between norms and facts makes sense and whether there can be expertise on values in the first place, this view of the divide between ethics and sociological analysis is not accurate. First, scholars in bioethics use refined empirical theories all the time, and they sometimes produce empirical knowledge in the course of their inquiries as well (although this is unlikely to be their main interest). In this case, for example, they ultimately make use of psychological theories pertaining to how human beings tend to deceive themselves when evaluating their interests against others, and hence need oversight if put in charge of potentially dangerous processes as experimentations with human subjects. Sociological analyses are in turn value-laden from the start, as many scholars would be ready to admit. Beliefs regarding what matters and what does not and the rankings of these different goods are crucial parts of the conceptual structure that let human beings navigate the complexity of reality. In this case, a key matter of concern that has influenced the inquiries of social analysts is public participation in science, both its potentialities for democratizing research and its misuses.

We can apply the same reasoning to a distinction that is often made, namely the distinction between "STS *in* bioethics" (e.g. Pickersgill 2013) and "STS *of* bioethics" (e.g. Evans 2006). While the latter would study bioethics as one among other objects of inquiry, the former supports ethical analyses by providing significant contextual information. In the current case a sociologist may want to reflect on how bioethicists have moved to become influential policy advisors and how their analyses turned into a tool of governance by limiting their inquiries to a very specific set of normative considerations and accepting the narratives that are put forward by the most powerful stakeholders. Bioethics is of course a legitimate object of study and indeed STS itself has been studied as a tool of governance suited to the new sensibility to public participation after the demise of technocratic command. However the boundaries are once again blurred: sociological analysis has employ refined, often implicit, normative theories to make meaningful claims pertaining to the role of bioethics in the governance of biotechnologies.

In the light of these continuities, it can be argued that a further way to conceive the relationship between STS and bioethics, to bridge the divide, is to conceptualize STS *as* a form of ethical analysis and viceversa. The two disciplines remain distinct in terms of histories, conceptual tools and methodologies. Their respective focuses, including which values and goods tend to highlight, are only partially overlapping. Bioethics, medical ethics and research ethics most often engage in moral balancing between individual and social harms and goods. STS is instead more attentive to issues of power, inclusion and democratization. The latter pays also much more attention to stakeholders´ goals, values and emotions, and STS scholars dedicate time to research these factors on the ground. We should not however take for granted the disciplinary boundaries as they are. These boundaries are after all the result of a highly contingent intellectual history that may not always track methodological divides that are insuperable or even valuable. The two disciplines offer accounts of the place of new technologies in society that are largely complementary, at least if we are not after decision algorithms, solutions or policy advices but we seek nuanced mapping of the stakes and the goods, and wider views of the practical options that are open to us. In summary, I would like to suggest relaxing the boundaries between the two disciplines. They both engage with particular sets of topics of human concern, try to understand their nuance and hopefully provide some guidance for action. For a thorough ethical analysis of research organization – and arguably of anything else – *the more the better*. Methodological and thematic pluralism is valuable. In discussions regarding how to make science better – discussions where STS scholars and bioethicists are hardly alone – we should cast our nets wide to catch as many values, goods and stakes as possible.

Acknowledgements This research was funded by the German Federal Ministry of Education, Research Grant 01GP1311.

References

Cooper, M. 2008. *Life as Surplus: Biotechnology & Capitalism in the Neoliberal Era*. Seattle: University of Washington Press.

Del Savio, L., A. Buyx, and B. Prainsack. 2016. Crowdsourcing the Human Gut: Is Crowdsourcing Also Citizen Science? *Journal of Science Communication* 15 (03): A03.

Dockser Marcus, A. 2014. The Ethics of Experimenting on Yourself. *Wall Street Journal*, October 24. Available at: http://www.wsj.com/articles/the-ethics-of-experimenting-on-yourself-1414170041. Accessed 24 Aug 2016.

Emanuel, E.J., D. Wendler, and C. Grady. 2000. What Makes Clinical Research Ethical? *JAMA* 283 (20): 2701–2711.

European Group on Ethics in Science and New Technologies. 2015. Opinion on the Ethical Implications of New Health Technologies and Citizen Participation. https://ec.europa.eu/research/ege/pdf/opinion-29_ege_executive-summary-recommendations.pdf. Accessed 24 Aug 2016.

Evans, J.H. 2006. Between Technocracy and Democratic Legitimation: A Proposed Compromise Position for Common Morality Public Bioethics. *Journal of Medicine & Philosophy* 31 (3): 213–234.

Graber, M.A., and A. Graber. 2013. Internet-Based Crowdsourcing and Research Ethics: The Case for IRB Review. *Journal of Medical Ethics* 39 (2): 115–118.

Hood, L., and S.H. Friend. 2011. Predictive, Personalized, Preventive, Participatory (P4) Cancer Medicine. *Nature Review of Clinical Oncology* 8 (3): 184–187.

Madeleine, A., and V. Rabeharisoa. 2012. Lay Expertise in Patient Organizations: An Instrument for Health Democracy. *Santé Publique* 24 (1): 69–74.

O'Connor, D. 2013. The Apomediated World: Regulating Research When Social Media Has Changed Research. *The Journal of Law and Medical Ethics* 41 (2): 470–483.

Physioprof. 2013. uBiome Has Made A Public Statement About IRB Compliance of Their Human Subjects Research. http://freethoughtblogs.com/physioprof/2013/02/21/ubiome-has-made-a-public-statement-about-irb-compliance-of-their-human-subjects-research/#ixzz3bzA69iGl. Accessed 24 Aug 2016. Read more: http://freethoughtblogs.com/physioprof/2013/02/21/ubiome-has-made-a-public-statement-about-irb-compliance-of-their-human-subjects-research/#ixzz4IGTruOHD.

Pickersgill, M.D. 2013. From 'Implications' to 'Dimensions': Science, Medicine and Ethics in Society. *Health Care Analysis* 21 (1): 31–42.

Prainsack, B. 2014. The Powers of Participatory Medicine. *PLoS Biology* 12 (4): e1001837.

Richman, J. 2013. Could a Citizen Scientist Win a Nobel Prize? http://www.tedmed.com/speakers/show?id=54370. Accessed 24 Aug 2016.

Sauermann, H., and C. Franzoni. 2015. Crowd Science User Contribution Patterns and Their Implications. *Proceedings of the National Academy of Science USA* 112 (3): 679–684.

Tempini, N., and L. Del Savio. 2018. Digital Orphans: Data Closure and Openness in Patient-Powered Networks. Fortchoming in *BioSocieties*.

Thorpe, C. 2010. Participation as Post-Fordist Politics: Demos, New Labour and Science Policy. *Minerva* 48 (4): 389–411.

Thorpe, C., and J. Gregory. 2010. Producing the Post-Fordist Public: The Political Economy of Public Engagement with Science. *Science as Culture* 19 (3): 273–301.

Tutton, R., and B. Prainsack. 2011. Enterprising or Altruistic Selves? Making Up Research Subjects in Genetics Research. *Sociology of Health and Illness* 33 (7): 1081–1095.

Vayena, E., and J. Tasioulas. 2013a. Adapting Standards: Ethical Oversight of Participant-Led Health Research. *PLoS Medicine* 10 (3): e1001402.

———. 2013b. The Ethics of Participant-Led Biomedical Research. *Nature Biotechnology* 31 (9): 786–787.

Vayena, E., R. Brownsword, S.J. Edwards, B. Greshake, J.P. Kahn, N. Ladher, et al. 2015. Research Led by Participants: A New Social Contract for a New Kind of Research. *The Journal of Medical Ethics* 42 (4): 216–219.

Wynne, B.E. 2006. Public Engagement as Means of Restoring Trust in Science? Hitting the Notes, but Missing the Music. *Community Genetics* 9 (3): 211–220.

Minding the Gaps: Sensitivities in Pursuing Empirical Ethics

Brian Rappert

1 Introduction

It is well established that policy agendas define and construct what counts as a concern. Yet, what remains outside of professional and policy agendas is also of considerable importance.

Why and how some topics are unrecognized, ignored, or not acted upon are matters amenable to social investigation and philosophical inquiry. This chapter sets out how sociologists and bioethicists can become more mindful about what they are *not* addressing. It does so by promoting a dialogue between sociology and philosophy (taken to include ethics for the purpose of this discussion); one that asks how they can learn from each other about their respective bounds. In terms of the overall aim of this volume to cross divides between disciplines, this chapter contends with how to navigate the fraught relation between 'ought' and the 'is'. While traditionally these two notions have been sharply distinguished and treated as legitimate topics for different disciplines (philosophy and ethics dealing with the former, and social sciences the latter), recent efforts are being made to hold the two together today under the label of 'empirical ethics'. These integrative efforts have proven contentious, in part because of the way they position how disciplines can contribute to each other. Rather than seeking to resolve disputes about whether and precisely what kind of 'empirical ethics' is needed, the goal of this chapter is to foster new sensitivities for approaching question about the relation between 'ought' and the 'is' – this by examining the theme of what is missing.

Towards this end, the next section offers some initial comments on the topic of this chapter and this analysis is then followed in the third section by disciplinary orientated commentary. As will be argued, what is not recognized, what is not

B. Rappert (✉)
Department of Sociology, Philosophy, and Anthropology, University of Exeter, Exeter, UK
e-mail: b.rappert@exeter.ac.uk

© Springer International Publishing AG, part of Springer Nature 2018   77
H. Riesch et al. (eds.), *Philosophies and Sociologies of Bioethics*,
https://doi.org/10.1007/978-3-319-92738-1_5

regarded as a concern, and what is not acted upon are *and* yet are not central concerns in sociology and bioethics. Attempts to study what is not taking place raise thorny issues such as the basis for identifying 'what's missing' as well as how notions of bias and ideology figure within academic analysis. With this sense of awareness, the third section details how attending to what is judged as absent could foster novel possibilities for crossing over divides.

2 Missing Matters

At any one time, only certain topics are prominent in public and professional discussions. What remains off the agenda can be judged as equally if not more significant than what is on it. This is evident in the way priorities vary by location and change over time. When student Jyoti Singh was beaten and raped in Delhi in 2012 (eventually resulting in her death), this case not only led to wide-ranging global attention to rape in India (at least for a time), it brought questioning regarding why Indian society and media had for so long turned a blind eye toward this crime as well as the misogynistic attitudes that had sanctioned it in the minds of some. At the time of the completion of this book, wide-ranging global attention is also being paid to misogyny and forms of consciousness and implicit bias against women. In these cases, as with others, with any newfound or heighted awareness of a topic then reasons for past indifference, apathy, or lack of regard can come to the fore. So too though can second order questions about how the time and energy dedicated to any one particular topic can have the effect of pushing others to the side.

Within the study of social and political life then, attention to what is blatantly observable, happening, etc. needs to be combined with what is not: what issues are not considered, what is not said, what decisions are not taken, what matters are rendered hidden, what grievance never get formed, what paths are never pursued.

And yet, with this eye towards what is not being considered comes a variety of issues. A critical set of questions that need to be posed is: For who, when, and in what manner are some matters 'non-matters'? Any newly identified trouble – such as how the latest advances in virology might enable ever more destructive weapons – is likely to be regarded in a myriad of ways across time and geography. This chapter takes as its object of consideration:

(a) for who, when, and in what manner are issues *not* recognized as posing concern;
(b) for who, when, and in what manner are issues *not* treated as a significant concern; and
(c) for who, when, and in what manner are issues are *not* acted upon.

In this sense, the absence of regard is treated as an irredeemably sociological phenomenon. As such, the descriptor of some matter as a 'non-issue' or 'missing topic' in the argument that follows should not be taken to refer to an absolute absence of consideration. Rather such descriptors are used as a shorthand for signalling relative absence vis-à-vis some individuals, some standard for what would count as sufficient attention, etc.

Let me offer a topic of my research. Under the code name "Project Coast", between 1981 and 1995 a chemical and biological warfare program was established and maintained in Apartheid South Africa. The investigations into this began in 1992 with a secret internal military inquiry into the involvement of covert military units in fomenting violence to undermine the process of transition to democracy. Later, in 1997, when the former leader for the program, Dr. Wouter Basson, was found in possession of large quantities of ecstasy the police narcotics unit started a separate investigation. Meanwhile the Office for Serious Economic Offences was investigating allegations of fraud made by scientists involved in the program. Another, separate investigation was undertaken by the Transvaal Attorney General (later the National Prosecuting Authority) into allegations of murder, torture, conspiracy and intimidation. Then, in 1997 the Truth and Reconciliation Commission set up to help move South Africa society from it fractured past began its investigation after having received applications for amnesty from scientists involved in the program. Finally, a professional inquiry was held by the Health Professions Council of South Africa to determine whether Basson was guilty of unprofessional and unethical conduct for activities undertaken as part of Project Coast.

These investigations gave rise to a three year criminal trial (that resulted in Basson's acquittal, or findings of not-guilty, on all charges); a public hearing of the Truth Commission and a report passing judgement; a monograph published by the United Nations Institute for Disarmament Research; another monograph based on the criminal trial; and a Health Professions Council of South Africa finding in 2013 that Basson was guilty of unethical and unprofessional conduct.

For some such a myriad listing of activities might be taken straightaway as justifying the conclusion that, yes indeed, Project Coast has been subject to substantial attention. And yet, grounds for a counter-evaluation can be offered too. Each investigation mentioned above to determine what happened has been structured and delimited by the very conditions that enabled it. Documentary traces and fragments complied to date signals much still remains unknown and perhaps will never be widely appreciated about what was done. As well, despite widespread public discussion about the project, its offensive intentions have never been officially acknowledged by South Africa or other nations – still at the time of writing (Rappert and Gould 2014). In short, many have found reason to call for the past to be left in the past. I and others have used these sorts of considerations to justify the need to further examine Project Coast within publications and funding proposals.

By what measure then, if at all, could one speak about Project Coast as a kind of unexplored, 'non-issue' in South Africa or elsewhere? Answers to this are likely to vary. In the case of the author, as with others, in effect the past investigations have acted as a stimulus for undertaking further lines of research and dialogue that raise the profile of the program. The purpose of much of this continuing effort has been to generate discussion about how the present is haunted though what is not known or unacknowledged about past (Gould et al. 2014). This stance is largely justified out of concerns for the state of transitional justice in South Africa, and in particular how suppression and secrecy still figure in the country today. Such an orientation though is just one basis that might be taken for characterizing what has been done and what might need doing regarding an encounter with the past. Which topics need what kind of attention though are ultimately deeply matters of politics.

The points raised in the previous paragraphs suggest the need for considered attention to how topics are positioned within academic claims making. Section 3 turns to this issue in relation to how disciplines position a sense of what is missing.

3 Disciplinary Engagements

Existing disciplinary agendas display a mixed and complex relation toward what is deemed absent. Take ethics. On the one hand, the case-based scenario reasoning prevalent in ethics is typically directed towards certain kinds of events: manifested dilemmas and choices. Furthermore, the charge has been levelled that this gaze is often "reactive" (to scandals, catastrophes, and so on), rather than pro-active to social priorities. For instance, agendas in bioethics have said to owe much to a fascination with the latest technologies, rather than the major public health problems measured in terms of the burden of disease (for instance, everyday infectious diseases – see Francis et al. 2005). As argued, this does not only result in slanted priorities, but a deficit in the applicability of core ethical notions such as informed consent and distributive justice.[1] And yet, while bioethical analysis is open to critique along these lines, it is a discipline that repeatedly calls into question what is taken for granted, deemed natural, or simply passed over.[2] To the extent bioethics as field was positioned as a response to the perceived limitations of mainstream ethics, it can be interpreted as born from a sense of the missing.

In a similar manner, while fields such as sociology are generally preoccupied with observing what is taking place, what is not taking place has also figured as a topic of study. As one example, for many decades attempts to understand the nature of power and domination sought to explain why some issues never garner enough attention to become treated as problems that need to be addressed through policy or others simply never get recognized (Bacrach and Baratz 1962). More generally, much of the research in sociology justifies itself as making what was only appreciated by those under study (whether coteries of elites or dispersed disenfranchised groups - Hess [2007]) more widely known. In this regard, many who study class, race, sexuality, and gender take 'giving voice' to problems sidelined by mainstream culture as their central aim (Bacchi 1999; Ryan-Flood and Gill 2010). Herein what should be given regard is set by what is commonly not. Lives forgotten, experiences

---

[1] See as well Tausig et al. (2006).

[2] See Fricker (2007) for an analysis of the ethical blind spots of social life as well as the discipline of ethics.

discounted, suffering ignored. As yet another area of example, in recent years renewed thought has been paid across many disciplines to the relation between knowledge, ignorance, and nonknowledge. As part of this, what is deemed 'unknown' need not be due to an insurmountable inability to find out. Instead, ignorance can be deliberately manufactured or result from an active choice by some not to know (Balmer 2012; Proctor and Schiebinger 2008). Uncertainty about the environmental and health repercussions associated with the use of certain pesticides, as an example, can result from a neglect of problems and a systemic failure to take heed of warning signals (Frickel et al. 2010; Gross and McGoey 2015).

Attention to what has not been considered has served as the basis for self-referential questioning of professional priorities in sociology. In the desire to give expression to experiences marginalized by mainstream culture, much of the sociological study of ethnicity in past decades in the West casted attention towards the social processes of constructing 'the Other'. More recently, scholarship has explicitly turned to the social production of 'whiteness'. In attending to how this often unacknowledged but pervasive collective identification is reproduced, such studies seek to question conventions in social life and social science. Just as new topics come into the fore over time, though, related ones can recede in the background. Today, some have argued that the relation between genetics, race and intelligence has become a kind of 'forbidden knowledge' among (at least American) sociologists due to its political potency (Kempner et al. 2011).

4   Empirical Ethics Revisited

The points made in the previous section indicate the complexity of what is at hand in this chapter. Whether a given matter is not recognized, not regarded as a concern, or not acted upon requires asking 'For whom?', 'When?', 'By what measure?', and so on. The distinction between what is judged as present and what is judged as absent is not so much a fuzzy boundary as it is a set of partial and competing images that fade or brighten depending on who is looking and for what purpose. That means a topic such as the international attention to the stockpiling of nuclear weapons (Brehm 2013) is likely to be characterized by such a shifting constellation of regard and disregard.[3] Moreover, within this landscape of regard, significantly different understandings of the 'same' issues are likely to be in circulation at any one time, thus suggesting the limits of perspectivist conceptualizations of social problems and underscoring the further need for a nuanced awareness.

The remainder of this chapter considers how attending to what is not being attended to could foster possibilities for undertaking disciplinary analysis as well as crossing disciplinary divides. It does so in relation to renewed calls to couple together ethics and social science under the heading of 'empirical ethics'. While this

---

[3] Moreover, the dormant status of a topic can result from the repeated failure over time from the recognition of a problem to muster concrete reform.

term is subject to contrasting definitions, it typically signals the need for greater multi-disciplinary investigation of ethical troubles. And while the originality of this call has not gone undisputed (Herrera 2008; Hurst 2010), much professional store has been placed in combining empirical methodologies for describing and explaining 'what is' with the scrutiny of 'what ought to be'.

Various typologies have been proposed to understand the relation of the 'ought' and the 'is', typologies that direct where a 'crossing' between ethics and social sciences needs to take place. Ives and Draper, for instance, offered distinctions based on the *purpose* of analysis (Ives and Draper 2009). Traditional 'philosophical bioethics' that engages in the specification of principles and logical argumentation for scholarly circles was contrasted with 'policy or practice oriented bioethics'. The latter was further divided into two approaches. A descriptive approach to policy works to depict reasoning in different situations while a normative approach theorizes what ought to be the case. For Ives and Draper it was only the latter where the relation of the 'ethics' to the 'empirical' needed to be addressed. Molewijk et al. (2004) outlined a five part typology of the uses of research data: (i) data for 'prescriptive applied ethicists' helps gauge compliance with theories of what ought to be; (ii) for 'the theorists' data can help rectify deficiencies in abstract moral reasoning; (iii) 'critical applied ethicists' engage in a back and forth refinement between empirical data and normative concepts; (iv) 'particularists' isolate the empirical study of what is taking place from ethical analysis; and finally as championed by Molewijk et al., (v) 'integrated empirical ethicists' seek to eliminate the divide between the factual and normative (and thus the divide between social science and ethics) by recognizing how facts depend on values and vice-versa (Molewijk et al. 2003).

In this section, rather than making a case for when, I want to concentrate on how the normative and empirical can be brought together. In their article, '"Nobody Tosses a Dwarf!" The Relation between the Empirical and the Normative Reexamined' Leget, Borry, and De Vries set out a five-stage process; one that seeks to retain a distinction between the ought and the is, rather than collapse the two together (Leget et al. 2009). In brief, the stages consist of:

1. *The determination of the problem*: Employing ethical principles and empirical research in tandem to ask what counts as a moral problem;
2. *The description of the problem*: Using normative concepts and principles as well as empirical information to unpack how problems are portrayed;
3. *Effects and alternatives*: Assessing the implications of possible choices by reference to consequences and concepts;
4. *The normative weighing*: Deriving judgments and recommendations about what should be done;
5. *The evaluation of the effects of a decision*: Subjecting decisions to normative and consequentialist appraisal.

I want to use this elaboration of stages to asks two questions: (i) 'How does attending to what is unrecognized, ignored, or not acted upon align with this five-stage model of empirical ethics?' and (ii) 'What alternative paths and sensitivities can help turn scholarly attention to what is unrecognized, ignored, or not acted upon?'.

As will be argued, through attending to what is not taking place, it is possible to think afresh about disciplinary preoccupations and divides.

4.1   The Determination of the Problem

In the first stage for Leget and colleagues, concepts in ethics mutually complement empirical investigation in the act of identification. Towards this end, ethics can 'sometimes perform the critical function of bringing to moral problems a sensitivity that is not felt by the majority of society's members' (ibid. 231). That can include helping 'empirical science discover implicit and otherwise invisible moral values present in their work' (ibid. 232). Empirical investigation can point out where normative work is needed as well as contribute to 'theoretical ethics by showing how ethical norms are embedded in culture and by revealing the ways society can conceal ethically questionable practices' (ibid. 232). An instance of this sort of contribution is given by Firth and colleagues. They sought to understand how those working in infertility clinics rendered their practice immune from the need for outside interference (Frith et al. 2011). The 'settled morality'[4] observed within the clinics was enabled by appeals to an abstract notion of ethos that notionally proscribed actions but in practice enabled ethical questions to be put to the side.

Such assigned tasks speak to the interweaving of what is recognized and what is not. Identifying that something should be given regard – whether through marshaling concepts such as dignity or data from surveys, interviews, etc. – is inseparable from the identification of what hitherto was not (but by some metric should have been) given regard. Indeed the main topic for the article by Leget et al. – dwarf tossing – represents both an identified problem and one not widely recognized. Just which one it is depends on for whom, when and how this judged.

As a result, in this first stage, setting about to identify what are not regarded as problems shares much with identifying what are so. A possible difference between the two emerges in the emphasis that the former would place on overcoming the limitations of both current concepts and empirical data. As envisioned by Leget et al., problem identification requires scrutinizing the moral values implicit within empirical study. Yet I what to argue that it is important to recognize that the scope for concern with the limitations of data gathering exercises is substantial; it extends far beyond the personal commitments of individual investigators (Kempner et al. 2011). Social science research is undertaken in circumstances that lead to the lack of recognition of some matters that, in turn, helps reproduce conditions of selective regard. Formal constraints on what can be studied, the structures of research organizations, and professional expectations about what is worth knowing are just some of the factors that delimit where attention gets directed which then also conditions the work of later researchers (ibid.).

---

[4] From Hoffmaster (1990).

In addition, in concerning ourselves with the absence of the recognition of problems, thought needs to be given to how the very procedures and routines designed to address ethical concerns can effectively render them unrecognized. As in the case of the introduction of euthanasia in the Netherlands, institutionalized clinical routines and policies informed by ethicists were said to be part of the de-problematizing of some acts (Houtepen 1998). Likewise, to treat patient care dilemmas in hospitals as matters of 'ethics' (rather than matters of professional power and authority) can reinforce relations of hierarchy in ways that result in the worries of some not being aired (Chambliss 1996). As another example, the backlash experienced by whistle-blowers can deter ethical issues from being pointed to in the first place (Martin 1999).

4.2   The Description of the Problem

In the second stage of the process as proposed by Leget et al., initial problem identification morphs into detailed description. Ethics and social research play respective roles here. The function of social research in this stage is two-fold: description and contextualization. With regard to description, empirical investigation should entail the 'careful and disinterested study of what is actually going on' (Leget et al. 2009). It is the job of 'theoretical ethics to make sure that researchers remain disinterested, pointing out when evaluative content enters their empirical descriptions' (ibid. 232). With regard to contextualization, 'social sciences can also perform a critical function by situating both ethical concepts and ethical problems in their broad social context' (ibid. 232). In the case of informed consent, for example, that might entail evidencing how and why decisions are made in concrete situations rather than how they are assumed to be made in theory.

When attending to the topics of this chapter, by contrast, empirical investigation entails the careful study of what is *not* going on. This alternative orientation provides a number of challenges for social research; challenges that problematize the tasks of description and contextualization as set out by Leget et al. One relates to open-endedness. The set of things that might be deemed as not taking place in any one situation is considerable if not overwhelming. Therefore, attending to what is not being attended to inevitably entails a radical culling down of possibilities.[5] Various research design proposals have been offered for identifying what absences should be the object of attention, but each demands close scrutiny (Dimitrov et al. 2007). Since this investigation relates to what is not happening, assessing the

---

[5] This is not to suggest that description and contextualization only become highly problematic in attempts to think about what is not taking place. Attempts to contextualize texts and objects entail a process of making connections, and how this done is open for contestation. Making connections can go on and on, without end, and along many paths (see Dilley 1999). In this sense, the manner in which this chapter seeks to open attention to the tensions and paradoxes of studying what is absent can be used a means for asking questions of inquiry more generally.

counterfactual relevance of potential factors at work is a demanding task, though past work would suggest it is not an impossible one. Strategies have included comparing shared pertinent similarities in order to account for the variations in how they are recognized, regarded or acted upon; utilizing quantitative research methods and non-cooperative game theory; providing historical background analysis; and undertaking structural analysis and probing individuals' perhaps initially unac-knowledged recognitions (see Gusfield 1981; Dimitrov et al. 2007; Eliasoph 1998). For instance, in a study of 'non-regimes' in international relations a group of authors took as their focus the lack of formal interstate policy agreements in situations that were favorable to regime formation. Such a tack though relies on analysts being able to determine when conditions are really favorable despite the lack of regime formation in practice.

The significance of attention to what problems are not given attention as well as the demands of doing so are evident from previous research that has sought to address the reasons for social quiescence. The lack of vocal expression of unease with any given situation need not reflect its absence. Instead it might reflect the lack of access to relevant information, an overabundance of information, a failure of commonplace concepts to map onto concerns, the dearth of places to voice shared experiences, etc. To go one step further, perhaps the most effectual exercise of dom-ination is to prevent concerns from ever being formed in the first place (Lukes 1974; Gaventa 1980).

The search for latent issues, in turn, brings additional demands for ethical sensi-tivity. Arguments seeking to explain the reasons for and the implications of how some wants, preferences, and grievances are never formed are likely to rely on nor-mative evaluations. As an example, individuals are often deemed responsible for inactions related to their social roles and capacities (for instance, why they did not 'blow the whistle'), but the degree to which this is the case is lessened if they make choices in line with structural constraints (for instance, if they did not denounce homophobic discrimination in the workplace when society at large was itself deeply homophobic – see as well Hayward 2006). For students of power though, social structure – however much the result of uncoordinated and unintended behavior – is still amenable to change because it is ultimately the result of actions and inactions. Thus, how determinations of responsibility are formed at the intersection of agency and structure is an important topic for normative analysis.

Attending to what is latent also provides reason to expand the appreciation of the issues at stake far beyond particular moments of 'decision making' as focused on in the Leget et al. model. One of those matters that can be addressed is how the need for 'decision making' is denied. Cunningham-Burely and Kerr (1999), for instance, suggested how geneticists fended off demands to address social and ethics troubles associated with their field. This involved combining the acquisition of an authority status that enabled them to secure funding and speak for the nature of the social consequences of genetics, while at the same time also distancing themselves from responsibility for addressing its negative consequences.

Acknowledgment of the import of what is not taking place does not lead to a straightforward design for empirical investigation. Simply relying on interviews or surveys to identity what is not being identified may well prove insufficient. In response, some social researchers have adopted the role of an authority figure who can specify what really should be of concern. Often that has entailed recognizing facts about others that they do not recognize in themselves (e.g., they are acting against their self-interests). Such a project though rests on two assumptions: (a) there exists something like 'real interests' or 'real grievances' and (b) that scholars of social life can discern them. Over the last several decades, much of the work in sociology, political science, and anthropology has distanced itself from such grand claims[6] (though appeals to putative interests still implicitly underpin much explanatory analysis[7]).

As an alternative way into describing problems, some have sought to make only modest and provisional claims about interests or to find alternative bases for grounding normative social science.[8] One strategy for the latter has been comparison: asking what has been of concern then but not now, there but not here, by them but not by us, and so on. In identifying mis-matches in how issues are regarded – for instance, why some identify security problems with synthetic biology and others do not (Evans and Frow 2015) – it is possible to posit that they should be given more attention.

Building on from the previous paragraph, striving to attend to what is not being attended to underscores both the demands and the likely limitations associated with Leget et al.'s suggestion that theoretical ethics could strive 'to make sure that researchers remain disinterested, pointing out when evaluative content enters their empirical descriptions' (Leget et al. 2009). The contingency and commitments of any single description come to the fore when it is looked at in terms of what it does *not* include. Through drawing attention toward certain directions and not others, descriptions of ethical problems necessarily provide selective and partial simplifications (Benford and Snow 2000; Gamson and Modigliania 1989). In this they frame what is going on and why, as well as prefigure what needs doing and by whom through what description include and what they do not include. The descriptions of social researchers and ethicists are not immune from concerns about what is not in them. The selective framing of issues is as pervasive and inevitable as it is disputable and contingent. As such, seeking to bring front and center topics that are deemed missing can be one way of putting these general conditions of analysis front and center.[9]

---

[6] See the 2006 special issue of Political Studies Review, volume 4 for a lengthy consideration of these and related points as well as Lukes (2005): Chapter 2.

[7] For an example, see Dimitrov et al. (2007).

[8] For instance, Hayward (2006).

[9] For a further discussion of related points see Woolgar and Pawluch (1985).

4.3   Effects and Alternatives

The third stage entails assessing the effects of actions and their alternatives. For Leget et al. (2009), empirical research can (i) 'overview of the consequences of a decision or policy' (ibid.: 233); (ii) 'test important aspects of normative arguments' (ibid.: 233) (such as whether the notion of 'slippery slope' is germane); and (iii) 'identify possible alternative solutions to moral problems' (ibid.: 233) through canvassing the views of experts and others.[10] For its part normative ethics can impart the relevance of deontological approaches and assess the casual relations at work in the depiction of ethical matters.

As with the previous two stages, in this one attending to what is not being attended to as advocated in this chapter helps alert us to the varied possible blinding effects of empirical investigation. For instance, matters can be rendered invisible for the purposes of decision making because data is gathered about one type of consequences over others (e.g., health studies that examine mortality levels over quality of life indicators – see Elliott [2013]). In this regard, ethical analysis has an important role to play in scrutinizing claims made by social researchers about the reasons why some topics become effectively off-limits. Research might be undone because of fears about the response to such research by an 'irrational' public (Kempner et al. 2011) or the priorities of funding agendas or the professional norms of researchers.[11] Going further, Rayner (2012) argues that some social research has had the effect of diverting and distracting societal and professional regard; resulting in certain matters not being acknowledged and some possibilities never being considered.

In the ways outlined above, empirical research can be part of the reasons why some matters are rendered unrecognized, not a concern, or not worth acting upon. In the effort to study what is absent then, the status quo and what foils efforts to move on from it demand consideration. To do this almost certainly requires looking beyond the consequences of a single policy or decision. What it ultimately requires by way of investigation though is much harder to specify.

4.4   The Normative Weighing

Stage four for Leget et al. involves drawing on the prior stages to determine what should be done. As such, it shares much with them for the purpose of this chapter. As with the previous stages, in stage four the empirical and the normative are tightly coupled. 'The most important contribution of theoretical ethics is to make sure that the normative power of the factual is put in its proper place' (Leget et al. 2009: 233).

---

[10] As an example of a work that strives for such contributions, see Haimes and Taylor (2011).

[11] For an example see Green (2005). For a wider discussion on 'undone science' see Frickel et al. (2010).

Empirical research should bring to 'light the values present in the wielding of moral ideas' (ibid.) by examining how bioethical weighings are being undertaken in practice. This mutual questioning is meant to reveal the underlining assumptions and the implicit commitments of ethicists and social researchers.

Taking what is absent as the object of consideration enables pointed questions to be asked of empirical data and ethical decision making. This is so because inquiring why information is not being gathered or something is not happening underscores the need to orientate to data and decisions as the effects of and affecting existing social practices, networks and routines.[12] Acknowledging indebtedness mitigates against treating either in isolation.

4.5   The Evaluation of the Effects of a Decision

As set out in Leget et al.'s account of the fifth stage, 'Once a decision has been made, it is the moral responsibility of the decision-makers to be sure that the decision has no unforeseen and unintended effects or side-effects. The critical function of theoretical ethics here is to make sure that the attention to effects does not, in the end, result in a victory of consequentialism (that is, if that is not the appropriate moral theory). Empirical researchers must map out the actual effects of a decision and provide data that can foster further ethical reflection' (Leget et al. 2009: 234).

By contrast, attending to what is not being attended to moves beyond decision making to instead consider the decisions not taken, not formalized, and not formulated as matters of concern. Such regard serves as a prompt for looking beyond this or that specific choice. What is called for is a way of questioning the contingencies of how some issues come to be defined as matters that warrant a decision in the first place as well as the consequences of the lack of decision making.

Approaching the discussion of ethical troubles in terms of what is not taking place also offers a corrective to only minding those matters that successfully garner priority. The spaces and opportunities afforded for raising issues by groups and organizations needs critical regard, least a false air of justification or complacency take hold.[13] This regard can also provide the basis for more rigorous empirically based and normative argumentation by helping to avoid dubious inferences that would result from only studying a subset of situations (such as studying only those matters that were subject to explicit decision making), encourage alternative hypotheses, and make a space for negative evidence (Dimitrov et al. 2007).

In its very complexity and open-endedness, examining what is missing invites appreciation of more than this or that choice. Focusing on such moments can be problematic because matters of ethics do not always come packaged in a discrete fashion. For at least some troubles, it can be the case that 'not only is the solution unknown,

---

[12] Indeed, it is the lack of efforts to look for evidence that is often at the center of political and regulatory disputes. For instance, see van Zwanenberg and Millstone (2005).

[13] For a consideration of this point see Winner (1992).

but the problem itself is initially not well defined, and the values that ought to drive its investigation and the valid methods to do so are unknown, unclear, or in dispute, as are the set of applicable theoretical models, the solution set, and the criteria for successful resolution' (Hoffmaster and Hooker 2009). Opening to what is missing highlights the need to avoid oversimplification of our understanding of what is going on. Any given matter of attention can entail a complex mix of (non-) recognition, (non-) concerns, and (non-)actions characterized by a tangled path of iterative movements of appreciation that do not have clearly demarked beginnings or endings (Majone 1989).

5 Discussion

This chapter has examined how the unrecognized, ignored, or not acted upon figures within the (i) agendas and activities of individuals, groups and organizations as well as (ii) the agendas and activities of social scientists and ethicists. As argued, rather than going out and probing straightforwardly overt, recognized issues, analysis can usefully cultivate regard for what is off the radar screen. Of vital importance in this pursuit is questioning the taken for granted assumptions about what counts as an ethical or social 'problem' in the first place.

As argued, this pursuit enables stretching questions to be posed about how to bring empirical and normative analysis together because such an undertaking requires addressing stretching questions for the social science and philosophy: How can they be attentive to what it has hitherto been ignored? How can claims that something is somehow absent be substantiated? How can commonplace preoccupations and priorities be overcome? (Rayner 2012). As such, inquiry into what is missing should not be thought of simply as an effort to fill in the holes of the existing stock of knowledge. Instead, it needs to be a project of questioning how the understandings of issues are formed and thereby what gets defined as missing, absent, unacknowledged, etc.

In relation to empirical ethics, this chapter has suggested that doing so potentially opens up possibilities for moving beyond those matters that commonly garner priority, for disturbing the status quo, for asking pointed questions about the values as well as the limitations of empirical data and ethical reasoning, and for becoming alert to the possible blinding effects of bringing facts to light. In doing to it is not only possible to critically and reflexively attend to the limits of social research and ethical analysis, it is possible to identify further ways in which disciplines can and need to come together in a skilful manner.

References

Bacchi, C.L. 1999. *Women, Policy and Politics*. London: Sage.
Bacrach, P., and M. Baratz. 1962. Two Faces of Power. *American Political Science Review* 57: 632–642.

Balmer, B. 2012. *Secrecy and Science: A Historical Sociology of Biological and Chemical Warfare.* London: Ashgate.

Benford, R.D., and D.A. Snow. 2000. Framing Processes and Social Movements. *Annual Review of Sociology* 26: 11–39.

Brehm, M. 2013. *Banning Nuclear Weapons.* London: Article 36.

Chambliss, D. 1996. *Beyond Caring: Hospitals, Nurses and the Social Organization of Ethics.* Chicago: University of Chicago Press.

Cunningham-Burely, S., and A. Kerr. 1999. Defining the "social". *Sociology of Health & Illness* 21 (5): 647–668.

Dilley, R. 1999. *The Problem of Context: Perspectives from Social Anthropology and Elsewhere.* Vol. 4. London: Berghahn.

Dimitrov, R., D. Sprinz, G. DiGusto, and A. Kelle. 2007. International Nonregimes: A Research Agenda. *International Studies Review* 9: 235–236.

Eliasoph, N. 1998. *Avoiding Politics.* Cambridge: Cambridge University Press.

Elliott, K.C. 2013. Selective Ignorance and Agricultural Research. *Science, Technology, & Human Values* 38 (3): 328–350.

Evans, S., and E. Frow. 2015. Taking Care in Synthetic Biology. In *Absence in Science, Security and Policy*, ed. B. Rappert and B. Balmer. London: Palgrave Macmillan.

Francis, L.P., M.P. Battin, J.A. Jacobson, C.B. Smith, and J. Botkin. 2005. How Infectious Diseases Got Left Out – And What This Omission Might Have Meant for Bioethics. *Bioethics* 19 (4): 307–322.

Frickel, S., S. Gibbon, J. Howard, J. Kempner, G. Ottinger, and D. Hess. 2010. Undone Science: Charting Social Movement and Civil Society Challenges to Dominant Scientific Practice. *Science, Technology and Human Values* 35 (4): 444–473.

Fricker, M. 2007. *Epistemic Injustice: Power and the Ethics of Knowing.* Oxford: Oxford University Press.

Frith, L., A. Jacoby, and M. Gabbay. 2011. Ethical Boundary-Work in the Infertility Clinic. *Sociology of Health & Illness* 33 (4): 570–585.

Gamson, W., and A. Modigliania. 1989. Media Discourse and Public Opinion on Nuclear Power. *American Journal of Sociology* 95 (1): 1–37.

Gaventa, J. 1980. *Power and Powerlessness.* Chicago: University of Illinois Press.

Gould, C., V. Harris, B. Rappert, and K. Smith. 2014. The Presence of the Past: Re-engaging the Legacy of South Africa's Chemical and Biological Warfare Programme. *Nelson Mandela Foundation News*, October 1.

Green, S. 2005. E3LSI Research: An Essential Element of Biodefense. *Biosecurity and Bioterrorism* 3 (2): 128–137.

Gross, M., and L. McGoey, eds. 2015. *Routledge International Handbook of Ignorance Studies.* London: Routledge.

Gusfield, J. 1981. *The Culture of Public Problems.* Chicago: University of Chicago Press.

Haimes, E., and K. Taylor. 2011. The Contributions of Empirical Evidence to Socio-ethical Debates on Fresh Embryo Donation for Human Embryonic Stem Cell Research. *Bioethics* 25 (6): 334–341.

Hayward, C.R. 2006. On Power and Responsibility. *Political Studies Review* 4 (2): 156–163.

Herrera, C. 2008. Is It Time for Bioethics to Go Empirical? *Bioethics* 22 (3): 137–146.

Hess, D. 2007. *Alternative Pathways in Science and Industry.* Cambridge: MIT Press.

Hoffmaster, B. 1990. Morality and the Social Sciences. In *Social Science Perspectives on Medical Ethics*, ed. G. Weisz. Boston: Kluwer Academic.

Hoffmaster, B., and C. Hooker. 2009. How Experience Confronts Ethics. *Bioethics* 23 (4): 214–225.

Houtepen, R. 1998. The Social Construction of Euthanasia and Medical Ethics in the Netherlands. In *Bioethics and Society*, ed. R. Devries and J. Subedi, 117–144. Upper Saddle River: Prentice Hall.

Hurst, S. 2010. What "Empirical Turn in Bioethics"? *Bioethics* 24 (8): 439–444.

Ives, J., and H. Draper. 2009. Appropriate Methodologies for Empirical Bioethics. *Bioethics* 23 (4): 249–258.

Kempner, J., J.F. Merz, and C.L. Bosk. 2011. Forbidden Knowledge: Public Controversy and the productIon of Nonknowledge. *Sociological Forum* 26 (3): 475–500.

Leget, C., P. Borry, and R. De Vries. 2009. "Nobody Tosses a Dwarf!" The Relation Between the Empirical and the Normative Reexamined. *Bioethics* 23 (4): 226–235.

Lukes, S. 1974. *Power: A Radical View*. London: Macmillan.

———. 2005. *Power: A Radical View*. Houndmills: Palgrave Macmillan.

Majone, G. 1989. *Evidence, Argument, and Persuasion in the Policy Process*. London: Yale University Press.

Martin, B. 1999. *The Whistleblower's Handbook: How to Be an Effective Resister*. Sydney: Envirobook.

Molewijk, B., A.M. Stiggelbout, W. Otten, H.M. Dupuis, and J. Kievit. 2003. Implicit Normativity in Evidence-Based Medicine. *Health Care Analysis* 11 (1): 69–92.

———. 2004. Empirical Data and Moral Theory. *Medicine, Health Care and Philosophy* 7 (1): 55–69.

Proctor, R., and L. Schiebinger, eds. 2008. *Agnotology*. Stanford: Stanford University Press.

Rappert, B., and C. Gould. 2014. Biological Weapons Convention: Confidence, the Prohibition and Learning from the Past. *ISS Occasional Paper*. 258 14 July 2014. Pretoria: Institute for Security Studies.

Rayner, S. 2012. Uncomfortable Knowledge. *Economy and Society* 41 (1): 107–125.

Ryan-Flood, R., and R. Gill. 2010. *Secrecy and Silence in the Research Process*. London: Routledge.

Tausig, M., M.J. Selgelid, S. Subedi, and J. Subedi. 2006. Taking Sociology Seriously. *Sociology of Health & Illness* 28 (6): 838–849.

Winner, L. 1992. Citizen Virtues in a Technological Order. *Inquiry* 35 (3/4): 341–361.

Woolgar, S., and D. Pawluch. 1985. Ontological Gerrymandering. *Social Problems* 32 (3): 214–227.

van Zwanenberg, P., and E. Millstone. 2005. *BSE: Risk, Science, and Governance*. Oxford: Oxford University Press.

'It's Not Just About Having Babies': A Socio-bioethical Exploration of Older Women's Experiences of Making Oncofertility Decisions in Britain

Alexis Paton

1 Introduction

Cancer and fertility have long been studied by a multitude of disciplinary perspectives. However women with cancer, and in particular older premenopausal women with cancer (i.e. above the age of 35, but still able to have babies), have often been ignored. In this chapter I discuss the experiences of this group of women when making fertility preservation decisions. I do so in order to develop awareness to this group of patients, but also to use their experiences to reflect on the policies, practices, and theories of decision-making that may impact the autonomy of this patient group. Patient autonomy is one of the pillars of good ethical practice, and thus it is vital that the bioethical theories and policies that dictate practice are properly scrutinised and updated with every new medical advancement.

Using the sociological analysis of empirical data to reflect on the accuracy of bioethical theories is still considered by some academics to be unorthodox. The use of sociological epistemologies to examine bioethical problems is part of what some have called the "empirical turn" in bioethics (Borry et al. 2005). This term, however, is something of a misnomer as, more often than not, the empirical turn only considers certain empirical methods as the only contribution sociology can make to bioethics (Paton 2017). Doing bioethics through the discipline of sociology is not just a passing trend, and the social sciences and its methods are not simple handmaidens for 'bioethics proper' to use for data collection (Haimes 2002; Paton 2017). Whilst the sociological examination of areas of bioethical interest have a long tradition (Haimes 2002; Paton 2017), it is only in the last 15 years that the field of bioethics has picked up its head and noticed just how much sociology is being done. Whether sociology should be part of bioethics has been extensively debated – two special

A. Paton (✉)
Social Science Applied to Healthcare Improvement Research, University of Leicester, Leicester, UK
e-mail: ahcp1@leicester.ac.uk

© Springer International Publishing AG, part of Springer Nature 2018   93
H. Riesch et al. (eds.), *Philosophies and Sociologies of Bioethics*,
https://doi.org/10.1007/978-3-319-92738-1_6

issues of *Bioethics* 2009, 23(4), and 2010, 24(5) provide a comprehensive snapshot of this debate. I will not, however, be engaging in this debate directly. Rather my sociological research is taken as a point of departure, one that shows sociology is a constitutive discipline of bioethics. As such, part of my aim is to demonstrate that sociological and social theory have a vital role to play in bioethical inquiry.

The research I describe here is situated in the field of medical sociology, but also within an ongoing debate about the role of bioethics and the constitutive fields of bioethics. I have previously argued for the importance of social and sociological theory in bioethics research (Paton 2017), and here I reiterate that argument, showing again how bioethics can benefit from the social sciences. By outlining how the research was conducted and analysed I put forth a case study for how bioethics can not only be "done" with sociological methods, but by using sociological and social theories; not just *methods* but *methodologies*.

I will begin by discussing why cancer and fertility, known as oncofertility, are an important area of bioethical exploration. I then outline how interviews were used to explore the way in which these patients make difficult oncofertility decisions. These interviews were used to examine existing social and bioethical theories of decision-making that inform current medical practice. While it was not the focus of the study to solely examine the experiences of older (aged 35–50) premenopausal women, they reported a qualitatively different experience when making decisions than those of the younger (aged 18–34) women who were also interviewed. In particular older women with cancer understood "fertility preservation" and "oncofertility techniques" to be useful tools for avoiding early menopause, and feel healthcare professionals' (HCPs) attitudes towards older premenopausal patients fertility needs may restrict their autonomy when making oncofertility decisions. By analysing the data using sociological and bioethical theories, I show how allowing for theoretical reflection through the social sciences may help to bring further understanding to the efficacy of existing theories of patient decision-making that currently inform practice.

2   Cancer and Fertility Preservation: A Brief Background

Cancer survival rates in Britain have risen to 50% and 1 in 20 of adult British women diagnosed with cancer each year are less than 50 years of age (Cancer Research UK 2014). As a result, the future fertility of cancer patients has become a critical consideration during treatment (Quinn et al. 2007). However, previous research has reported that cancer patients of all genders and ages feel under-informed about the effects of cancer treatment on their future fertility (Quinn et al. 2007; see also Schover 1999; Schover et al. 2002a, b; Letourneau et al. 2012; Peddie et al. 2012; Loren et al. 2013; Corney and Swinglehurst 2014). In particular there have been very few studies documenting the experiences of adult female patients, and studies on British adult female patients remain notably under-represented (Crawshaw et al. 2009).

Recent studies have begun to fill this gap, with Peddie et al. (2012) reporting that female cancer patients in Scotland valued the opportunity to learn more about fertility preservation (FP), however full access to this information remained difficult to achieve.[1] In addition, preserving fertility for childbearing purposes after cancer is one of the foremost reasons identified as to why FP techniques are so important to patients, highlighting the importance of FP to patients' perceptions of their life after cancer (Wilkes et al. 2010).

Reflecting the recent literature guidance from the National Institute for Health and Care Excellence, the British Fertility Society, the Royal College of Obstetricians and Gynaecologists, and the American Society for Reproductive Medicine, all recommend that healthcare professionals (HCPs) discuss the effect of cancer treatment on future fertility with patients (British Fertility Society 2003; R.C.P. 2007; Loren et al. 2013; N.I.C.E. 2013). Despite these recommendations there remains no national funding policy for FP in Britain (Wilkes et al. 2010), and the provision of current fertility services offered by the National Health Service (NHS) is unevenly spread across the country (Wilkes et al. 2010; Haimes and Taylor 2013). Recent studies have called for more steps to be taken to ensure female patients understand the extent to which their fertility may be damaged, and are aware of available FP techniques (Goodwin et al. 2007; Crawshaw et al. 2009; Wilkes et al. 2010; Letourneau et al. 2012; Peddie et al. 2012; Loren et al. 2013; Corney and Swinglehurst 2014). However, while life after cancer is increasingly becoming an important part of cancer treatment discussions, FP for cancer patients, in particular women, remains under-utilised in Britain (Wilkes et al. 2010).

3 Conducting a Sociological and Bioethical Research Project

So how does one "do" bioethics "with" sociology? The answer is: quite simply. The theoretical framework of this research was based in a sociological bioethics (Paton 2017), which prioritises sociological methods, methodologies and theory to collect empirical data to interrogate existing bioethical theory (Hoffmaster 2001; Haimes 2002; Hedgecoe 2004; Borry et al. 2005; Haimes and Williams 2007; Frith 2012; Paton 2017). Taking a decidedly sociological approach, a one-to-one interview-based, qualitative study was designed so as to access the lived experiences of the participants (Mason 2002; Silverman 2004). The research received ethical approval from both the university and local NHS research ethics committees. Interviewees aged 18–50 at the time of their diagnosis were recruited via cancer support groups throughout Britain. Participants were recruited through presentations made by the author at support group meetings, as well as pamphlets and flyers left with support groups, and information about the project was posted on social media sites. A project website was also set up for participants to view more details about the project at

---

[1] Corney and Swinglehurst's (2014) research adds to the growing body of literature on the experiences of British female patients, arguing that these patients require more detailed information on FP, and an in-depth explanation of whether they, as individual patients, are eligible or not to attempt FP before cancer treatment.

their leisure. Individuals who indicated an interest in participating were sent an information pack with a detailed explanation of the research, contact information and a consent form. Once participants had signed and returned the consent form, the participant was contacted by the author to arrange an interview.

In total eleven interviews took place. All were with women who were aged 18–50 at the time of their cancer diagnosis. Six of the participants were aged 35–50 at their diagnosis (i.e. older premenopausal women), and it is their interviews that are the focus of this chapter. No specific type of cancer was targeted during recruitment. All the interviews were conducted by the author. The interviews were audio recorded and transcribed verbatim with permission from the interviewees.

As with the data collection, the analysis of the data was also decidedly sociological. A grounded theory approach was employed, using coding, memo-writing, sampling for theory development, and constant comparison of the data during the analysis (Charmaz 2006). Data analysis that is conducted under the umbrella of interpretivist methodology is often done using a grounded theory approach, however rarely in its "pure form" (Brewer 2000: 108). As Brewer points out "there are very few genuine cases of grounded theory" (Brewer 2000: 109). Grounded theory-inspired analysis is often found in interpretivist research (as is the case here) because it adheres to the data and has a number of steps that can be used to help organise and understand the data (Brewer 2000). A 'pure' grounded theory could not be used due to time restrictions imposed on the research project, so thematic analysis based on salient or recurrent issues from within the interviews was also employed to help form the major thematic categories of the research (Attride-Stirling 2001).

Three major themes were identified from the research: "Time", "Information and Understanding", and "Being Guided". From within these themes emerged a subsidiary theme of the older premenopausal cancer patient's experience. This chapter draws on data from all three themes to highlight this sub-theme of the older patients' experiences.

Once categorised, the data was also analysed with reference to current social and bioethical theories of patient decision-making, examining whether these theories were reflected in current practice, as described by the patients' experiences. The bioethical theories of traditional autonomy (Beauchamp and Childress 2009) and relational autonomy (Sherwin 1992, 1998; Mackenzie and Stoljar 2000) were drawn on as these theories inform much of current policy on informed consent. Theory from medical sociology, in particular shared decision-making (Emanuel and Emanuel 1992; Charles et al. 1997, 1998, 1999; Annandale 1998; Nettleton 2013), was also used to analyse the experiences of the interviewees.

4   Premenopausal Cancer Patients' Experiences of Oncofertility

The interviews with the older women revealed their unique understanding of the terms "fertility" and "fertility preservation". Interviewees often used these terms to refer specifically to the preservation of their ovarian function (regardless of whether that was an accurate understanding of the terms). In particular FP was understood by the interviewees as a possible way to avoid the onset of early menopause after cancer treatment. The interviewees understanding of the term "fertility" and the role of FP in cancer treatment manifested in two themes that emerged from the interviews: 'Avoiding early menopause' and 'HCP assumptions about who wants FP'.

Avoiding early menopause will be discussed first as the women's concerns about early menopause after cancer were often felt to be misunderstood by their HCPs, which led to the second theme that emerged in the interviews.

4.1 Avoiding Early Menopause

The desire to avoid early menopause was often something that was realised in retrospect. All the older women interviewed had entered early menopause as a result of their cancer treatment. The interviewees expressed frustration at not being fully informed, or, in some cases not being informed at all, about the likelihood of entering the menopause early. When discussing FP techniques in the interviews the women understood these techniques as possible options that might have helped them avoid their early menopause. In many cases this was a misunderstanding of what FP offers, however it did not change the importance that FP had for the women. Avoiding the menopause was desirable for all the women interviewed as they felt that early menopause would be (and was) a difficult burden to bear for their physical and mental wellbeing post-cancer.

The physical toll the menopause had on their bodies was a particular concern for the older women interviewed. This was acutely apparent in the participant Angela's[2] experience. Angela was a breast cancer survivor with a family history of ovarian cancer. As she was at a very high risk of developing it, Angela had been advised to have her ovaries removed as a preventive measure against ovarian cancer. Despite those risks Angela explains why she felt strongly about not losing her ovarian function, which she understood as her "fertility":

> ...I'd hung on to my ovaries deliberately because I wanted to maintain them...I thought [fertility] was good...keep up the bone density...I just felt that I wanted to hold on to my ovaries as much as possible.

Angela describes the maintaining of her ovarian function as a "good" thing. Because of this perceived good in remaining premenopausal Angela attached quite a lot of importance to keeping her ovaries, even in the face of mounting evidence that she may very well have ovarian cancer. The older premenopausal women interviewed valued their ovarian function so highly that they were often willing to take risks with their health in order to maintain that function for as long as possible. This was not always possible, and in some cases the women were not made aware of the effects of early menopause on their bodies. Mary was unaware of the full impact of going into

---

[2] All participant names and identifying details have been changed to protect anonymity.

the menopause early, and expressed frustration at how little is communicated to patients about the menopause and the physical effect this has on patients post-cancer:

> Probably the one thing which doesn't get mentioned very much is in connection with the early menopause…is the impact on sexual activity…in my experience that has been by far the biggest impact and you know the most unpleasant one…because it's a sensitive issue… it tends to get brushed aside…and it is quite a big one and can last for a long while after the cancer treatment is gone.

As Mary points out, the arrival of the menopause is the longest lasting symptom of cancer treatment for these women, and it affects the patient's life after cancer regardless of remission or survival. Another participant, Robyn, reiterated this concern, expressing a point shared by all the participants by saying that the impact of the menopause on the patient's life is rarely discussed when the decision to proceed with potentially fertility damaging cancer treatments is made:

> I also hadn't been prepared for the impact of early menopause, that was quite difficult for my body to take…It was quite hard. Simple things like hot flushes are a major issue even now.

In addition to concerns about the impact of early menopause on their physical health, the women also expressed their concerns about the effect they felt early menopause had on their mental wellbeing. These concerns were in addition to any effects that having cancer may have also had. For example, Anne was concerned about how the menopause would affect her overall mood:

> I think that one of the things was how removing what they were going to take out was going to affect me…would that mean they would take the ovaries and what hormonal effects that would have on like mood. Obviously like to do with your sexual relationships as well. Because you know you hear stories about menopause and things like that.

Monica was similarly concerned about how her mental wellbeing would be affected if she agreed to the double mastectomy and hysterectomy (and resulting early menopause) recommended by her doctors:

> It just makes you feel like there is nothing left that makes you a woman. Nothing that defines you as a woman. Everything that is textbook defines you as a woman is gone. To get the menopause as well is just icing on the cake!

Finally, Diane summed up her feelings as "sad" about not having the choice to preserve fertility or avoid early menopause:

> Yeah it's weird but [losing fertility is] connected to it and I do feel sad that I had to go into the menopause early.

All of the older women interviewed were keen to avoid losing their ovarian function and remain premenopausal. None of them felt that they were near enough to their natural menopause to feel comfortable entering it early. Two of them were actively considering having more children, and all the women were concerned about the effects of early menopause on their wellbeing. These concerns reflect how highly the women valued their continued ovarian function, and their resulting reactions to entering early menopause highlight the distress they felt in having to endure an unwanted, possibly avoidable, physical consequence of their cancer treatment. This distress was compounded for many of the women as they felt they were not adequately prepared for early menopause by their HCPs, or given the option of avoiding early menopause using FP techniques. The older women felt that had they been fully aware of the consequences of cancer treatment to their ovarian function and fertility they perhaps would have made different choices about those treatments.

In expressing desires to retain their fertility and ovarian function it is possible that these women were also expressing desires to retain their femininity. Perhaps. Though few interviewees explicitly discussed what menopause had done to their understanding of their selves as women and what it means to "be" a woman, feminine and femininity. Additionally, whether gender performativity played a role in these women's desires to use fertility preservation is unknown, as there is little evidence of that in the interviews. Instead, the focus, as can be seen in the quotes above, is on biology and the role fertility (and its constituent working parts) play in maintaining a premenopausal life without hot flushes, osteoporosis and vaginal dryness (to pick up on interviewees' articulated concerns). To discuss the role that femininity and gender performance may have had on the interviewees' decisions is conjecture at best, and for this reason I would like to touch on these possibilities in this section without engaging in analysis of these concepts in this chapter.

4.2   HCP Assumptions About Who Wants Fertility Preservation

Those interviewed as part of this project felt that concerns about their future fertility, how they valued continued ovarian function, and how they understood the role FP could play in maintaining ovarian function, were often brushed aside by their HCPs. This was perceived to be the result of assumptions about their age or a misunderstanding of how patients valued their continued fertility. These assumptions were often inconsistent with the beliefs and informational needs of the patient, and HCPs limited the information about FP techniques that was provided. Angela expresses this concern when she explains how chemotherapy was presented to her by her HCPs:

> But the subject of fertility…Nothing was said at first…I'd hung on to my ovaries deliberately…and [the doctor] said, "Well chemo is going to stop your ovaries working…and that will be it"…nobody said "Do you want to hang on to your fertility"…nothing was actually mentioned.

Due to these mistaken assumptions Angela was unaware of the FP options available to her when she was quite keen to preserve her fertility (and ovarian function) for as long as possible:

> …I would have wanted someone to make a bigger effort to talk about fertility, not just say "Oh, this is going to happen"… I would have seriously wanted someone to talk about saving some eggs somewhere along the line…I would have wanted someone to save my fertility…

A similar situation occurred with Robyn, who was an older first time mother. Robyn felt her HCP made assumptions that she had no interest in preserving her fertility:

> … but there wasn't really any kind of discussion about whether… I wanted more children… So it was probably assumed that with my age and everything that I wouldn't want children any way… So whether [fertility] was something that maybe should have been discussed in a little more detail, I'm guessing that it probably should have been.

Like all the other women interviewed, Robyn felt strongly that all aspects of her treatment, including the effect of cancer treatment on her fertility and possible ways to preserve her ovarian function, should have been discussed. Many of the interviewees did not want more children, however they were concerned about preserving their ovarian function; a concern they felt was not always shared by their HCPs. All of the interviewees said they would have liked to have heard about FP options regardless of their age and family size. The interviewees also felt that the assumptions made by the HCPs interfered with their decision-making by limiting the options that were made available to them.

5   Bioethics and the Oncofertility of Older Women with Cancer

In this study women aged 35 or more understood "fertility" and "fertility preservation" to be synonymous with "ovarian function preservation" and "avoiding the menopause." The interviewees wanted more information and discussion than they in fact received about their available options for the preservation of fertility/ovarian function. Having this information, and the opportunity to discuss it, was important to interviewees as they viewed it as a necessary component of their informed decision-making. Interviewees felt assumptions made by their HCPs limited the information available to them as patients, information which may have changed the decisions they took about their treatment. This opinion was most often voiced with reference to the information they were given about the effects of chemotherapy on fertility, and the availability of FP techniques.

These findings add to previous research that suggests there is an ongoing and persistent breakdown in communication between cancer patients and their HCPs (Peddie et al. 2012; Loren et al. 2013; Corney and Swinglehurst 2014; Letourneau et al. 2012), one that is evidently difficult to repair. Patients are consistently reported as asking for better, more detailed discussions with their HCPs about their disease and treatment options to better facilitate decision-making in the medical context (Crawshaw et al. 2009; Wilkes et al. 2010). Studies on decision-making outside the oncofertility context show similar findings (Street et al. 2007; McMullen 2012; Zikmund-Fisher et al. 2012; Frongillo et al. 2013; Kim et al. 2013).

The need to address these concerns about the quality of HCP-patient discussions about FP is important, as HCP perceptions and assumptions have a significant influence on the information communicated between patient and HCP (Cope 2002; Street et al. 2007; Quinn and Vadaparampil 2009). Several studies have reported that HCP beliefs about who should be informed of oncofertility techniques restrict the information that patients receive (Cope 2002; Nisker et al. 2006; Siminoff et al. 2006; Goodwin et al. 2007; Quinn and Vadaparampil 2009; Peddie et al. 2012). These studies report that none of the criteria used to determine if FP techniques should be offered were related to the patient's own perceived fertility needs. Instead HCPs are less likely to discuss FP options if the patients are older, single, in a perceived unstable financial situation, identify as homosexual, or are HIV positive (Cope 2002; Nisker et al. 2006; Goodwin et al. 2007; Quinn and Vadaparampil 2009; Peddie et al. 2012).

Feminist perspectives on choice in the healthcare context have also highlighted how the assumptions made by HCPs are especially damaging for women making medical decisions. Such assumptions further confine women's already limited options, and thus restrict their autonomy in the medical encounter (Mackenzie and Stoljar 2000). As communication with their HCPs is one of the key ways that patients achieve an informed understanding of their disease and treatment options (Kim et al. 2013), the lack of communication demonstrates the detrimental effect that these assumptions can have on informed, autonomous patient decision-making.

This information is highly valued by patients. By holding back information they think would be irrelevant, HCPs may be forcing patients to make decisions from a restricted or incomplete set of options. How information is communicated, and the quality of that information, are essential components of two key sociological and bioethical concepts that underpin the contemporary understanding of patient decision-making: shared decision-making and patient autonomy. Both concepts argue that informed understanding is a necessary part of decision-making for patients. Patients should not have to make a decision until that informed understanding has been achieved (Mackenzie and Stoljar 2000; Beauchamp and Childress 2009) however in practice informed understanding is not safeguarded or ensured in a way that reflects the weight it is given in theory (Anspach 1993; Corrigan 2003; McMullen 2012). The experiences of the women who participated in this study show how difficult it can be to achieve even a minimal level of informed understanding in high-pressure areas of medicine. Where a patient's life may hang in the balance and where decisions have to be made quickly, such as is often the case in oncology, HCPs may make erroneous judgement calls about what information to present to their patient, so that the patient can make an informed decision.

Problematically both the research on decision-making in oncofertility, and in the wider clinical context, continue to suggest that the solution to the 'difficult' decision-making that occurs in practice can be mitigated by more comprehensive and accurate discussions about the management of fertility (Peddie et al. 2012; Zikmund-Fisher et al. 2012; Loren et al. 2013; Nettleton 2013; Corney and Swinglehurst 2014). This suggestion ignores not only the empirical evidence that comprehensive and accurate discussion is difficult to achieve in the current medical encounter, but also that those theories that inform current practice, namely shared decision-making and traditional accounts of patient autonomy, fail to accurately reflect how patients make decisions, either idealised or in practice (Hedgecoe 2004; McMullen 2012).

Examining the views of older premenopausal women highlight the persistent challenges to patient decision-making in the clinical encounter. The experiences of the older women in this study show how patients are still obligated to make decisions within difficult to achieve decision models that operate under restrictive circumstances, and ignore the patient's own values, beliefs and informational needs. These experiences reinforce the growing evidence that decision-making in the clinical encounter is not only informed by outdated theories, but that current policies informed by those theories fail even to adhere to them in practice.

It is not appropriate to offer policy recommendation here; that is not the goal of this chapter, nor is this possible given the limited number of patient interviews (Whittemore et al. 2001; Morse et al. 2002). Instead these findings suggest that further research with both patients and clinicians is needed to develop a more complete picture of how decisions are (and should be) made in the oncofertility context in Britain. However, if patients and HCPs can engage in more effective, reciprocal communication, the potentially negative influence of HCP assumptions could be minimised. I have previously argued that in order to achieve the type of communication necessary to uphold patient autonomy in the clinical context, further work must be done to introduce practices that mitigate the persistent power imbalance between patient and doctor (Paton 2017). Without these power imbalances, assumptions about who wants FP may be more possible, and discussion about treatments and the preservation of fertility/ovarian function may be more comprehensive, and tailored to the patient's informational needs, thus fulfilling the theoretical requirements for informed understanding and facilitating patient autonomy (Paton 2017). The experiences of the participants of this study provide further evidence that the policies in place in the NHS with regard to patient decision-making require further research and reflection such that the practices informed by the policies are reflective of current experiences, patient needs and theoretical obligations (Paton 2017).

6   Conclusion

One of the most interesting findings of the overall study was what 'fertility preservation' meant to the women interviewed. For the interviewees, knowing about FP techniques was not just about 'having babies'. The women in this study understood 'fertility' to go beyond childbearing; preserving fertility was about choice preservation for the future, and in particular, not going into early menopause after cancer. Research on oncofertility that only focuses on patients who wish to preserve fertility for childbearing reasons might therefore overlook an oncofertility patient-base who want to preserve their ovarian function. This preference is important to consider as it is a key component of how older women understand FP techniques, and constitutes a part of their informed understanding that they use to make autonomous decisions. Future social science research is needed to examine how older pre-menopausal women, and their HCPs, experience cancer and fertility decisions, as well as how fertility is understood and valued by women with cancer who do not want to have children. This research, and research like it, can also be used to further our understanding of the efficacy of existing bioethical theories which inform policies and practices around patient decision-making in the clinical context. While bioethics has been predominantly developed through the fields of philosophy, law and theology, it also has a strong sociological tradition that is now coming to the forefront (Paton 2017). Research like that described in this chapter, shows not only how much the field of bioethics can benefit from sociological work, but how necessary that work is to developing bioethics in such a way that it is current, accurate and efficient.

Declaration of Interest   The author reports no conflicts of interest. The author alone is responsible for the content and writing of the paper.

References

Annandale, E. 1998. *The Sociology of Health and Medicine: A Critical Introduction*. Cambridge: Polity Press.

Anspach, R.R. (1993). Deciding Who Lives: Fateful Choices in the Intensive Care Nursery. Berkeley: University of California Press.

Attride-Stirling, J. 2001. Thematic Networks: An Analytical Tool for Qualitative Research. *Qualitative Research* 1: 385–405.

Beauchamp, T.L., and J.F. Childress. 2009. *Principles of Biomedical Ethics*. 6th ed. Oxford: Oxford University Press.

Borry, P., P. Shotsmans, and K. Dierickx. 2005. The Birth of the Empirical Turn in Bioethics. *Bioethics* 19 (1): 49–71.

Brewer, J. 2000. *Ethnography*. Buckingham: Open University Press.

British Fertility Society. 2003. A Strategy for Fertility Services for Survivors of Childhood Cancer. *Human Fertility* 6: A1–A40.

Cancer Research UK. 2014. *All Cancers Combined Key Facts*. Retrieved from http://www.cancer-researchuk.org/cancer-info/cancerstats/keyfacts/Allcancerscombined/. Accessed 8 Sept 2014.

Charles, C., A. Gafni, and T. Whelan. 1997. Shared Decision-Making in the Medical Encounter: What Does It Mean? (or It Takes At Least Two to Tango). *Social Science and Medicine* 44 (5): 681–692.

Charles, C., C. Redko, T. Whelan, A. Gafni, and L. Reyno. 1998. Doing Nothing Is No Choice: Lay Constructions of Treatment Decision-Making Among Women with Early-Stage Breast Cancer. *Sociology of Health & Illness* 20 (1): 71–95.

Charles, C., A. Gafni, and T. Whelan. 1999. Decision-Making in the Physician-Patient Encounter: Revisiting the Shared Treatment Decision-Making Model. *Social Science and Medicine* 49: 651–661.

Charmaz, K. 2006. *Constructing grounded theory: A practical guide through qualitative analysis*. Thousand Oaks: Sage Publications.

Corney, R.H., and A.J. Swinglehurst. 2014. Young childless women with breast cancer in the UK: a qualitative study of their fertility-related experiences, options, and the information given by health professionals. *Psycho-Oncology* 23: 20–26.

Cope, D. 2002. Patients' and Physicians' Experiences with Sperm Banking and Infertility Issues Related to Cancer Treatment. *Clinical Journal of Oncology Nursing* 6 (5): 293–295.

Corrigan, O. 2003. Empty Ethics: The Problem with Informed Consent. *Sociology of Health & Illness* 25: 768–792.

Crawshaw, M., A. Glaser, J. Hale, and P. Sloper. 2009. Male and Female Experiences of Having Fertility Matters Raised Alongside a Cancer Diagnosis During the Teenage and Young Adult Years. *European Journal of Cancer Care* 18: 381–390.

Emanuel, E.J., and L.L. Emanuel. 1992. Four Models of the Physician-Patient Relationship. *The Journal of the American Medical Association* 267 (16): 2221–2229.

Frith, L. 2012. Symbiotic Empirical Ethics: A Practical Methodology. *Bioethics* 26 (4): 198–206.

Frongillo, M., S. Feibelmann, J. Belkora, C. Lee, and K. Sepucha. 2013. Is There Shared Decision Making When the Provider Makes a Recommendation? *Patient Education and Counselling* 90: 69–73.

Goodwin, T., B.E. Oosterhuis, M. Kiernan, M.M. Hudson, and G.V. Dahl. 2007. Attitudes and Practices of Pediatric Oncology Providers Regarding Fertility Issues. *Pediatric Blood Cancer* 48: 80–85.

Haimes, E. 2002. What Can the Social Sciences Contribute to the Study of Ethics? Theoretical, Empirical and Substantive Considerations. *Bioethics* 16 (2): 89–113.

Haimes, E., and K. Taylor. 2013. What Is the Role of Reduced IVF Fees in Persuading Women to Volunteer to Provide Eggs for Research? Insights from IVF Patients Volunteering to a UK 'Egg Sharing for Research' Scheme. *Human Fertility* 16 (4): 246–251.

Haimes, E., and R. Williams. 2007. Sociology, Ethics and the Priority of the Particular: Learning from a Case Study of Genetic Deliberation. *British Journal of Sociology* 58 (3): 457–476.

Hedgecoe, A. 2004. Critical Bioethics: Beyond the Social Science Critique of Applied Ethics. *Bioethics* 18 (2): 120–143.

Hoffmaster, B., ed. 2001. *Bioethics in social context*. Philadelphia: Temple University Press.

Kim, J., A.M. Deal, U. Balthazar, L.A. Kondapalli, C. Gracia, and J.E. Mersereau. 2013. Fertility Preservation Consultation for Women with Cancer: Are We Helping Patients Make High-Quality Decisions? *Reproductive Biomedicine Online* 27: 96–103.

Letourneau, J.M., E.E. Ebbel, P.P. Katz, K.H. Oktay, C.E. McCulloch, W.Z. Ai, A.J. Chien, M.E. Melisko, M.I. Cedars, and M.P. Rosen. 2012. Acute Ovarian Failure Underestimates Age-Specific Reproductive Impairment for Young Women Undergoing Chemotherapy for Cancer. *Cancer* 118 (7): 1933–1939.

Loren, A.W., P.B. Mangu, L.N. Beck, L. Brennan, A.J. Magdalinski, A.H. Partridge, G. Quinn, W.H. Wallace, and K. Oktay. 2013. Fertility Preservation for Patients with Cancer: American Society of Clinical Oncology Clinical Practice Guidelines Update. *Journal of Clinical Oncology* 31 (19): 2500–2510.

Mackenzie, C., and N. Stoljar. 2000. *Relational Autonomy: Feminist Perspectives on Autonomy, Agency and the Social Self*. New York: Oxford University Press.

Mason, J. 2002. *Qualitative Researching*. 2nd ed. London: Sage Publications.

McMullen, L.M. 2012. Discourses of Influence and Autonomy in Physicians' Accounts of Treatment Decision Making for Depression. *Qualitative Health Research* 22 (2): 238–249.

Morse, J.M., M. Barret, M. Mayan, K. Olsen, and J. Spiers. 2002. Verification Strategies for Establishing Reliability and Validity in Qualitative Research. *International Journal of Qualitative Methods* 1 (2): 13–22.

National Institute for Health and Care Excellence. 2013. *Fertility: Assessment and Treatment for People with Fertility Problems*. Retrieved from http://www.nice.org.uk/guidance/cg156/chapter/recommendations#people-with-cancer-who-wish-to-preserve-fertility. Accessed 8 Sept 2014.

Nettleton, S. 2013. *The Sociology of Health & Illness*. 3rd ed. Cambridge: Polity Press.

Nisker, J., F. Baylis, and C. McLeod. 2006. Choice in Fertility Preservation in Girls and Adolescent Women with Cancer. *Supplement to Cancer* 107 (7): 1686–1689.

Paton, A. 2017. No Longer "Handmaiden": The Role of Social and Sociological Theory in Bioethics. *IJFAB* 10 (1): 30–49.

Peddie, V., M. Porter, R. Barbour, D. Culligan, G. MacDonald, D. King, J. Horn, and S. Bhattacharya. 2012. Factors Affecting Decision Making About Fertility Preservation After Cancer Diagnosis: A Qualitative Study. *BJOG* 119: 1049–1057.

Quinn, G.P., and S.T. Vadaparampil. 2009. Fertility Preservation and Adolescent/Young Adult Cancer Patients: Physician Communication Challenges. *Journal of Adolescent Health* 44: 394–400.

Quinn, G.P., S.T. Vadaparampil, C.K. Gwede, C.A. Miree, L.M. King, H. Clayton, C. Wilson, and P. Munster. 2007. Discussion of Fertility Preservation with Newly Diagnosed Patients: Oncologists' Views. *Journal of Cancer Survivorship* 1: 146–155.

Royal College of Physicians, The Royal College of Radiologists, Royal College of Obstetricians and Gynaecologists. 2007. *The Effects of Cancer Treatment on Reproductive Functions: Guidance on Management* (Report of a Working Party). London: Royal College of Physicians.

Schover, L.R. 1999. Psychological Aspects of Infertility and Decisions About Reproduction in Young Cancer Survivors: A Review. *Medical and Pediatric Oncology* 33: 53–59.

Schover, L.R., K. Brey, A. Lichtin, L.I. Lipshultz, and S. Jeha. 2002a. Knowledge and Experience Regarding Cancer, Infertility, and Sperm Banking in Younger Male Survivors. *Journal of Clinical Oncology* 20 (7): 1880–1889.

———. 2002b. Oncologists' Attitudes and Practices Regarding Banking Sperm Before Cancer Treatment. *Journal of Clinical Oncology* 20 (7): 1890–1897.

Sherwin, S. 1992. *No Longer Patient: Feminist Ethics and Health Care*. Philadelphia: Temple University Press.

———. 1998. *A Relational Approach to Autonomy in Health Care*. Philadelphia: Temple University Press.

Silverman, D. 2004. *Qualitative Research: Theory, Method and Practice*. London: Sage Publications.

Siminoff, L.A., G.C. Graham, and N.H. Gordon. 2006. Cancer Communication Patterns and the Influence of Patient Characteristics: Disparities in Information-Giving and Affective Behaviours. *Patient Education and Counseling* 62: 355–360.

Street, R.L., Jr., H. Gordon, and P. Haidet. 2007. Physicians' Communication and Perceptions of Patients: Is It How They Look, How They Talk, or Is It Just the Doctor? *Social Science and Medicine* 65: 587–598.

Whittemore, R., S.K. Chase, and C.L. Mandle. 2001. Validity in Qualitative Research. *Qualitative Health Research* 11 (4): 522–537.

Wilkes, S., S. Coulson, A. Crosland, G. Rubin, and J. Stewart. 2010. Experience of Fertility Preservation Among Younger People Diagnosed with Cancer. *Human Fertility* 13 (3): 151–158.

Zikmund-Fisher, B., M.P. Couper, and A. Fagerlin. 2012. Disparities in Patient Reports of Communications to Inform Decision Making in the DECISIONS survey. *Patient Education and Counseling* 87: 198–205.

'Can Someone Please Decide?' How the Media Represent the Risk of Drinking During Pregnancy

Hauke Riesch

1 Introduction

The nature of scientific and, in particular, biomedical research is that it is beset with uncertainties. When it comes to advising individuals, health advice is rarely clear-cut. Rather, it is often qualified with statistical statements. As a result a fair amount of attention has been paid to the communication of risk and the effects and content of uncertainty (cf. Brashers 2001). Furthermore, because the media plays a large role in disseminating scientific findings, a lot of research has concentrated on the way such uncertainties are represented (Jensen 2008; Stocking and Holstein 2006).

One particular type of uncertainty is the conflicting messages that can arise from expert disagreement and, therefore, the need to interpret conflicting health advice. This can take on a particular salience for the public, one that it might not have for the experts involved in the relevant debate, and who will be more certain of their own background knowledge and assumptions. Furthermore, as Funtowicz and Ravetz's concept of post-normal science indicates, even when we are confronted with essentially the same information, there are often different but equally legitimate perspectives to be taken on risk situations (Funtowicz and Ravetz 1993). Thus, conflicting advice on health matters and their representation in the media is an active area of research (e.g. Friedman et al. 1996).

This chapter examines conflicting perspectives on the risks of drinking during pregnancy. I do so by using Rose's (1992) "prevention paradox" which provides one way to examine why it is that the same risk can mean different things to different people. Since the construction of uncertainty and its representation in the media is a complex matter, a detailed case study analysis of one controversy and the media coverage that resulted was chosen to extract and identify more subtle features in the media coverage of health risks. A recent episode, where health experts from two

H. Riesch (✉)
Department of Social and Political Sciences, Brunel University London, London, UK
e-mail: Hauke.riesch@brunel.ac.uk

© Springer International Publishing AG, part of Springer Nature 2018
H. Riesch et al. (eds.), *Philosophies and Sociologies of Bioethics*,
https://doi.org/10.1007/978-3-319-92738-1_7

different public health bodies disagreed over the precise wording of their advice, is taken as this chapter's focus. It is used to present a case study in the lay interpretation of risk as well as the dynamics of conflicting expert advice. While the actual difference between the sets of advice may have been rather slight, the disagreement publicly highlighted the uncertainty inherent in medical research. It also occasioned heated debate within the newspapers about public health advice and its motivations.

The case study in this paper concerns a brief episode in 2007–2008 when two UK public health bodies appeared to give conflicting advice about the consumption alcohol during pregnancy and the way in which this was reported by the national press. Between May 2007 and March 2008 the Department of Health (DoH) and the National Institute for Health and Clinical Excellence (NICE) were updating their advice booklets for antenatal care. Although both agencies had previously advised that consuming low amounts of alcohol during pregnancy was fine, in May 2007 the DoH altered its position and advised complete abstinence. On the other hand NICE broadly maintained their position and, when they published their draft revised guidelines in October YEAR 2007, they continued to advise women that the consumption of small amounts of alcohol was not a concern. This paper will look closely at the two sets of advice issued by the DoH and NICE. This will be compared with the interpretation of this advice in the associated media reports, and the ensuing opinion and editorial (op-ed) coverage offered by the UK national press. In the light of a quite fierce press reaction to what was, in effect, a rather minor discrepancy in the advice issued by the two bodies, this chapter will finish with some observations on the difficulties associated with issuing public health advice in contexts where the perspective offered by the prevention paradox has particular salience; on topics where the scientific evidence is not yet clear and the individual risks are low.

One of the particular issues addressed by this paper is the "prevention paradox" which was introduced to the field of preventive medicine by Geoffrey Rose (1992). Getting people to change lifestyle habits in order to prevent possible future health risks is a complex undertaking and beset with difficult problems, not least insofar as it involves persuading individuals that further reducing risks that are already rather low will be a direct and personal benefit to them. As Rose puts it, a "large number of people exposed to a small risk may generate many more cases than a small number of people exposed to a high risk" (1992: 24). Therefore behaviours that present an important public health risk from the perspective of the Department of Health or similar may, from the point of view of the individual, appear negligible. The debate in the case study presented in this chapter exemplifies the issue identified in the quotation from Rose, and this perspective helps explain why there were such strongly held differences of opinions concerning the advice from different experts. Although the 'prevention paradox' has become a frequently cited insight within the discipline of preventive medicine, it has not often been featured in explanations of why perspectives on risks differ within risk communication and media studies. This paper adds to the research on media risk presentation by introducing Rose's paradox within a broader sociological setting.

In addition to this chapters two main themes – conflicting health advice and the prevention paradox – I develop and present my analysis in such a way that it can be applied more widely to public health advice. I do so by discussing some of the other factors that affect this case study. These include the actual scientific evidence underlying the advice, the moral and ethical considerations that pertain to the issue, and, finally, the effect of a strict 'no-drinking' message on the population as a whole. These three themes will be addressed in the background section.

2   Risk and the Media

The way people learn about risks of lifestyle behaviour is often through a combination of official sources and the news media covering health stories. In particular, the media's role in informing the public about particular risks has become a central part of risk research (Kitzinger 1999; Allan 2002). However Kitzinger cautions that we cannot automatically tell from media coverage what the readers' reactions are to risk topics. Wahlberg and Sjöberg's (2000) review of newspaper influences on their readership showed that it is not quite as influential as often supposed. Similarly, Lupton and Chapman (1995) found that readers can be very cynical about press coverage of risk. Most studies of risk in the media focus on more general examinations of representations of a particular risk over some specified time period. For example Washer (2004) has examined the media reaction to SARS, Weingart et al. (2000) focus on the global warming debate in Germany, whilst Kitzinger and Williams (2005) have attended to stem cell research. More relevantly for this paper, Connolly-Ahern and Broadway (2008) have considered media representations of Foetal Alcohol Spectrum Disorder (FASD). In their work they have reviewed the last 10 years of media coverage of FASD in the US, finding that the media discussion on the topic often gave conflicting advice. This means that there is a "pressing public relations need within the governmental organizations responsible for combating FASD" (Connolly-Ahern and Broadway 2008: 381).

One of the features of news reporting of contentious scientific issues is the issue of balance. Generally speaking media outlets try to give a fair account of the issue and often do so by giving a similar amount of time to opposing viewpoints. This can result in stories about scientific research appearing to be more uncertain than they really are; media reporting often fails to accurately convey the degree to which there is a consensus amongst scientists. Boykoff and Boykoff (2004), for example, suggest this is the case in reporting of global warming. However, Stocking (1999) notes that the contrary claim can also be made; that journalists present provisional and uncertain results as more certain than they are (for example on marginal health benefits of new drugs).

There is also a moral dimension in much official and media risk discourse. Following the work of Douglas, Lupton (1993) argues that that risk discourse has taken on a distinctively moral hue. This is especially true in the context of health advice, where putting oneself at risk through lifestyle choices is increasingly seen as morally dubious. According to the discourse of lifestyle risks:

> If individuals chose to ignore health risks, they are placing themselves in danger of illness, disability and disease, which removes them from a useful role in society and incurs costs upon the public purse. Should individuals directly expose others to harm – for example by smoking in a public place, driving while drunk, or spreading an infectious disease – there is even greater potential for placing the community at risk. (Lupton 1993: 397)

The moral element of the discourse on health risks is not only evident in policy discourse about health and lifestyle, but in discussions of treatment for those whose lifestyle contributes or 'causes; their illnesses, particularly in media coverage (Brown et al. 1996). Other examples of how public health and lifestyle issues are moralized (or maybe conversely how moral issues are medicalized) include obesity (Inthorn and Boyce 2010), smoking (Rozin and Singh 1999) and, more recently, vaping (Tamimi, The Ethical Framework for the Use of E-Cigarettes, this volume).

By taking the approach of looking closely look at a particular controversy that in itself is only a very small part of a much larger story, this paper analyses the wording of official advice and its interpretations in detail, focusing on how these helped trigger a public debate about the issue. It will also identify some of the inherent difficulties for public health bodies in advising on lifestyle risks.

This kind of close textual examination ought to be understood in relation to larger-scale analyses of media stories about health risks, and also of the socio-historical development of the pregnancy-alcohol risk debates. Whilst apparently resting on little actual evidence, such issues have been widely and publicly debated, So much so, that it has even been analysed as a "moral panic" (Armstrong and Abel 2000). This study will supplement the perspective advanced by this literature by looking at the British, rather than the North American context. In so doing it will show how this topic has a particular dynamic, one which resists identifying simple actors such as the media, the public, or the government as the main drivers behind it. Golden (2005) and Armstrong (2003) frame debates over FAS as arising out of the US cultural background, especially surrounding abstinence and the foetal rights movement. This paper will contribute to the social analysis of FAS by having a specific look at how the arguments are debated within the British context, where FAS has recently moved prominently into policy debates (Lowe and Lee 2010), but so far received little sociological attention.

3   Background to the Debates on Drinking During Pregnancy

3.1   Fetal Alcohol Spectrum Disorder

First described by Jones and Smith (1973), Foetal Alcohol Syndrome (FAS) is characterised by physical, behavioural and cognitive abnormalities. FAS can therefore have debilitating effects on the child's life. Less severe conditions have often been called "Foetal Alcohol Effect" (FES) leading to the view that consuming alcohol in pregnancy can result in a spectrum of harmful effects on children, and the idea of "Fetal Alcohol Spectrum Disorder" (FASD). Full FAS is nevertheless notoriously difficult to diagnose. Furthermore, it is often not a desirable to make a diagnosis in the absence of a treatment (Armstrong 2003: 126). Estimates about the prevalence of FAS and especially FASD vary greatly, and it is not clear that alcohol is the exclusive cause. The American Academy of Pediatrics (AAP 2000) quotes a prevalence of 5.2 per 10,000 births in the US, with higher rates for some subgroups such as Native Americans (30 in 10,000). Certainly, the population with the greater number of cases of FAS and FASD display unhealthier drinking patterns during pregnancy, but there are also higher rates of other conditions such as poverty, smoking addiction and malnutrition (Armstrong 2003). In addition, even among heavy alcohol users FAS occurs in only about 5% of births (Armstrong cites Abel 1995). FAS is not, then, a foregone conclusion even among those who clearly abuse alcohol during pregnancy. Indeed, FAS can sometimes develop in mothers who are not consistently heavy drinkers. It seems the type of alcohol consumed and the manner of consumption may also play a role, with binge drinking considered to be more harmful than drinking a comparable amount across a period of a week (Abel 1999).

There is, then, a considerable amount of uncertainty about the prevalence of FAS and its precise cause(s). It seems reasonably certain that low and steady amounts of alcohol during pregnancy pose very low risks to the foetus. But this also means that there is no particular safe limit under which we can say with absolute certainty that it is safe to drink. Though this has led many medical researchers to advocate complete abstinence (for example AAP 2000), many of the physicians interviewed by Armstrong (2003: chapter 4) argued that there is no harm in drinking a little.

FAS and FASD are the main health problem associated with drinking during pregnancy. While there are other concerns over drinking in pregnancy, notably a higher probability of miscarriage especially during the first trimester (Harlap and Shiono 1980; Kesmodel et al. 2002), it is FAS (and FASD) which has mostly coloured the health advice on the topic as well as the subsequent public discussion.

3.2   Alcohol, Pregnancy and Morality

Early on in the course of the debate, public discussion of FAS took on a distinctly moral element, particularly in the United States (Armstrong 2003). In her analysis of the history of the public debates around FAS, Golden (2005) distinguishes between the different and distinct frames in which the issue has been debated:

> Over the course of three decades FAS became first a diagnosis, then a public health problem, and next a morality tale about mothers. Now FAS would also be identified as an "abuse excuse," emblematic of the public's concern about individual responsibility and moral order, and thus it would be, in some settings, demedicalized. (pp. 152–153)

The way one understands FAS, and the stance one takes in or on the debate, is coloured by other background beliefs and assumptions. For example, Golden argues that those who generally consider alcoholism to be an illness, as opposed to reflecting an individual's moral failings, directly impacts on whether FAS is perceived as an unfortunate event or as a form of child abuse.

Social scientific analysis of this issue has so far been largely focused on North America. Connolly-Ahern and Broadway's (2008) recent media frame analysis of FASD relied on papers from the Northeastern United States, and Golden's (2005) analysis of the history surrounding the public debates on FAS covers only the US context. Armstrong (2003: 205) even specifically states that the "United States seems exceptional when we view it in an international context," particularly when one considers that the incidence of FAS is slightly higher in the US, while drinking rates are slightly lower, than is the case in comparable European countries. At the time Armstrong was writing, only a few European countries advised complete abstinence like the US general surgeon (Armstong 2003: 205). Even then the phrasing of the advice was not as strong. There are, however, signs that a seemingly similar official attitude is starting to take root in Europe. For example, at around the same time that the story in this paper took place, France joined the US in introducing legislation requiring warning signs to be displayed on alcoholic drinks (Institut National de Prévention et d'Éducation pour la Santé 2007). In Britain a tightening of official attitudes towards drinking while pregnant and the resulting risk of FAS has also been apparent (Lowe and Lee 2010).

As it does not relate to the mother putting her own life at risk but, potentially, risking the health and wellbeing of an unborn child, risk in relation to pregnancy in general is a very emotional and often moralised issue. Ruhl has analysed risk discourse in popular advice manuals, arguing that "pregnancy is increasingly portrayed as a state requiring careful and detailed risk prevention" (1999: 95). She argues that a "risk model of pregnancy" has come to dominate. This model emphasises "the individual responsibility incumbent upon the pregnant woman to provide her foetus with the best possible gestational environment." (Ruhl: 102). The issue of alcohol during pregnancy has a particular interest beyond mere lifestyle risks, because the two issues that intersect here both have very emotional backgrounds. Furthermore, alcohol itself is often seen as a dubious pleasure, whilst female drinking has, historically, always had a special significance (Thom 1994); continuing nowadays with media criticism of "ladette culture" (Day et al. 2004).

On a more general level, the importance placed on lifestyle risk factors can be analysed as a reconnection of personal responsibility for what, in many ways, are chance events. The assumption that adopting a healthy lifestyle means we can avoid the risk of ill health, is often translated in to its converse: meaning that we often assume that someone's ill health is a product of the person not leading a healthy lifestyle. If that is the case then we are, to an extent at least, being rendered responsible for our own illnesses. As it is perceived as a very dispensable pleasure alcohol features prominently in this regard.

In the British popular press, debates about alcohol and government advice run along two main concerns. More often than not, alcohol is perceived to be a problem. Particular examples include teenage binge drinking, national embarrassment at the behaviour of British holidaymakers in foreign resorts, and for turning Friday night city centres into no-go areas for older people. Such concerns have a distinct class association and alcohol is rarely seen as a middle class problem. When health agencies flag up concerns over middle class drinking, the press reaction is overwhelmingly negative (Riesch and Spiegelhalter 2011), adopting the stance that the state has no business telling us how to lead our lives. This is a frequent trope in British newspaper coverage of public health advice (often dismissively referred to as being representative of the "Nanny State"). Concerns over FAS clearly raise different concerns than other examples of problematic drinking habits do, not least because expecting mothers do not easily embody the role of the caricatured drunken teenager, holiday maker or weekend partygoer.

3.3 Advice and Educational Interventions

There is little agreement about how best to advise pregnant women over alcohol consumption. It can either be merely to provide information and leave the decision to the individual or to discourage drinking to the point of abstinence. Neither is it the case that there is much consensus over how best to achieve the aim once it is settled on. As a consequence it is unsurprising that two different pieces of advice on the same issue can sound very different, even if they start from the same scientific and ethical viewpoints.

Blanket abstinence advice such as that given by the US Surgeon General has featured heavily in many campaigns to encourage pregnant women to drink less alcohol. There are, however, considerable doubts over what they can actually achieve. As has been pointed out in both media and professional literature, it is unlikely for example that the women most at risk, such as alcoholics, will be swayed by such warning labels, while the ones who are already worried about leading a healthy lifestyle anyway are the ones who will most likely be influenced by them. There are many variations and nuances in the advice that public health bodies give on this topic, and they vary not only between different countries and cultures, but even within the same country (O'Leary et al. 2007).

The next section will present a narrative, beginning around May 2007 and continuing to March 2008, about how two public health bodies have tightened their advice on pregnancy and alcohol, and how these changes were covered in the UK national press. The analysis is based on documents taken from the websites of NICE and the Department of Health, and the UK national press. The newspaper articles were sourced from a search on the online newspaper database LexisNexis. The search terms were for "NICE" (or spelled out) and "alcohol" and "pregnancy" in the national UK press between March 2007 and May 2008. This was supplemented by searches for "alcohol" around the dates when the reports and draft reports came out. Articles on alcohol that were not part of this story were filtered out. The newspapers searched were all those UK national papers covered by the LexisNexis database: these were the broadsheets Times and Daily Telegraph, Guardian and Independent. The tabloid press covered was the Daily Mail, and the red-top tabloids the Sun, the

Daily Mirror and the Daily Star. All these papers were joined by their respective Sunday sister papers including the now defunct News of the World. The search resulted in a total of 76 articles.

The chapter follows the "risk story" approach that David Spiegelhalter and I have used to follow the evolution of the interpretation of risks from science to press release to newspapers (Riesch and Spiegelhalter 2011), and which Lee et al. (2016) applied to a similar recent case involving a study on the fetal alcohol exposure and its effects on children's IQ.

4 Findings

4.1 Conflicting Advice

In October 2007 several newspapers in the UK reported on a draft for revised guidelines on pregnancy which were to be published by NICE in March 2008. The guidelines were broadly similar to previous guidelines from 2003 which advised that during pregnancy women should limit themselves to no more than one (UK) unit a day. This became newsworthy primarily because it contradicted advice published by the Department of Health earlier in 2007 that pregnant women should not drink any alcohol at all. Because these were only draft guidelines, the story made the newspapers again in March 2008, when the final version was published, this time the advice was changed to conform more with the DoH advice. This section will provide a chronological account of how the story developed.

The original advice from NICE in the 2003 guidelines was that

> [e]xcess alcohol can harm your unborn baby. If you do drink while you are pregnant, it is better to limit yourself to one standard unit of alcohol a day (roughly the equivalent of a small glass of wine, a half pint of beer, cider or lager, or a single measure of spirits). (NICE 2003: 118)

When the guidelines were reviewed in 2007 this advice was essentially retained, not least because no further substantial evidence about drinking and pregnancy had since come to light. There were, however, some minor changes. In October 2007 the draft guidelines read:

> Pregnant women should limit their alcohol intake to less than one standard drink (1.5 UK units or 12g of alcohol) per day and if possible avoid alcohol in the first 3 months of pregnancy.

> Women should be informed that binge drinking (defined as more than 5 standard drinks on a single occasion) may be particularly harmful during pregnancy. (NICE 2007a: 36)

One change was the addition of a sentence on binge drinking being particularly harmful, a point which had not previously been stated. Another was the inclusion of the warning that women in the first trimester should, if possible, not drink at all. In this regard, then the new advice can be considered stronger than the previous formulation. However it was also slightly less strict: They advised to abstain "if possible", which suggests that certain circumstances may lead to the advice being overridden. In addition, the NICE guidelines talk about standard drinks rather than units of alcohol. A standard drink is usually more than one unit of alcohol. However, rather than advising women should "limit yourself to" one unit the advice states that should consume "less than" one drink.

By the time the draft NICE guidelines were officially published, the Department of Health had already updated their own advice (in their "pregnancy book"), half a year previously. This states that

> The UK's Chief Medical Officers advise that, as a general rule, pregnant women, or women trying to conceive should avoid drinking alcohol. If you do choose to drink, to protect your baby you should not drink more than one or two 'units' of alcohol once or twice a week and should not get drunk. (DoH 2007: 14)

The difference between these two statements was noted by the press and resulted in the publication of several news articles which focused on the discrepancy. However, the changes made to the NICE advice were of emphasis more than content: Although abstinence was advised as a general rule, the 1 unit a day limit was repeated for women who still want to drink alcohol. In fact, compared to the revised Nice guidelines, the above passage from the DoH pregnancy book could be seen as slightly less strict. It does not mention that particular attention should be paid during the first trimester. Nevertheless, the DoH still talked about one unit per day, rather than standard drink (which is a bit more than a unit), and the limits it sets are therefore lower than those offered by NICE.

4.2   The Take-Up of the Story in the Press

The new pregnancy book from the Department of Health already attracted newspaper attention when it came out in May 2007, even though the advice was not particularly focussed on abstinence, at least as compared to the previous editions. However, it is interesting to compare the reaction to the 2007 edition of the DoH's pregnancy book to how the NICE guidelines were to be received in October of that year. For example, op-eds in the Guardian and the Independent argued that the revised DoH advice of abstinence is patronising to responsible women, and that the advice was only really aimed at "the 9% of women who still drink above the previous recommended levels" (Curtis, Guardian 25 May 2007). Furthermore, emphasis was placed on a statement from a spokesperson from the department of health. This suggested that the rules were not based on any new evidence, but that instead the new stricter wording was adopted to avoid confusion (Shaikh, Independent, 25 May 2007). A few days later, the Daily Mail criticised the guidelines for being part of the "Nanny State", arguing that the people at whom the advice is aimed at, "teenage binge-drinkers", will be "too bladdered to read any of the scary statistics anyway." ("Why our Nanny State Should Keep Mum," Daily Mail, 30 May 2007).

Half a year later, the draft revised NICE guidelines became available, and most newspapers reported on the conflict between it and the DoH advice from May. This attracted much more attention than the previous release of the pregnancy book, and invariably the difference between the two sets of advice was emphasised. Thus the Guardian's headline was "Confusion over advice on alcohol for pregnant women". This is the Guardian's interpretation of NICE's advice:

> In practice, this would mean the green light for women to drink one small glass of wine a day, or half a pint of 5% lager or strong cider, or a bottle of alcopop. (Boseley, Guardian, 11 Oct. 2007)

The Times similarly wrote on the conflict between the two guidelines, with the headline "Pregnant women told glass of wine a day is fine – and too dangerous". Again, though the paper tried to explain the two sets of advice and the reasoning behind them, the actual wording got interpreted in ways that highlight the differences between them.

> The experts [at NICE] concluded women should avoid alcohol only during the first trimester, but said drinking led to a slightly higher risk of miscarriage. (Rose, Times, 11 Oct. 2007)

In fact, NICE actually wrote that the evidence of a higher risk of miscarriage for low levels of drinking is "limited and of poor quality" (p. 108), a point which the Times also acknowledged. There is, however, a difference in emphasis. The low levels of confidence that NICE has about the evidence is reflected in its less cautionary advice, while the Times' interpretation suggested that NICE had given the advice despite the evidence (however slight), and therefore painted a picture of NICE as being less concerned about potential miscarriage. The Times also quoted the British Medical Association, which explained the more precautionary approach taken by the DoH guidelines.

The Daily Mail also focused on the confusion. The reporting followed a similar structure to the other newspapers quoted above, tough this article was different from the other ones in that it also included a quote on the reasoning behind the NICE guidelines:

> [Dr David Williams] said: 'I think the NICE advice is accurate according to the data we have got – a total ban is not a good thing. Most women who think about it will try to avoid alcohol in pregnancy but if they are really missing it, a glass of wine now and again is a good tonic to the spirits. (MacRae, Daily Mail, 11 Oct. 2007)

Finally, the Daily Telegraph published a short but front-paged article on the affair. Although it pointed towards the conflict in the advice given by NICE and the DoH, it also explained the reasoning behind the DoH's stronger wording. It highlighted the fact that the DoH guidelines themselves had recently been changed

> amid fears that women were either ignoring advice that they could consume a little, had various ideas of what the recommended amount was, or were being given different advice in different parts of the country. (Clout, Daily Telegraph, 11 Oct. 2007)

Thus, even though the Telegraph wrote about the contradiction between the guidelines and possibly accentuating it as much as the other papers, it did at least cover more of the reasoning behind the stronger formulation.

Although the story was not extensively covered in the op-ed sections, a Daily Mail editorial a few days later continued the theme from May of casting doubts on the abstinence advice:

> It still won't be illegal to drink during pregnancy, for sure, but I can't help wondering how long before this new initiative gives a licence to every tutting busybody to harass and humiliate pregnant women enjoying a night out. [...] I don't believe for a minute that alcohol is good for developing foetuses, but I reckon the jury's still out on the harm that a moderate amount can actually do. ("Here's Another Quasi-Scientific Solution to Make Mothers Feel Bad When Things go Wrong...," Daily Mail, 16 Oct. 2007)

The national debate about the issue of FAS was taken up by the British Medical Journal, which a couple of days later published a "Head to Head" feature (for and against abstinence advice in pregnancy). Advocating the earlier advice and arguing that a little alcohol is fine O'Brian argued, among other points, that

> the strong advice not to drink implies a certainty and confidence in the evidence that simply does not exist. There is a danger that our stance is perceived as paternalistic and will lead to a loss of confidence in medical advice in general. (O'Brian 2007)

Arguing for the abstinence advice, Nathanson et al. (2007) wrote that only a strong message can avoid causing women confusion, particularly when one considers that most do not have a good idea of what a unit is, and that the evidence for harm is inconclusive but not ruled out.

This debate shows that there is still a fair amount of disagreement in the expert community, this includes disagreement over how to best proceed on giving advice, even if nobody disagrees about the evidence at hand. The two articles were reported on by several newspapers in a way that betrays editorial preoccupation, but in both cases they highlight the apparent confusion over what pregnant women are allowed to drink. The Independent chose to highlight O'Brian's contribution (Headline: "Pregnant women 'could drink'", Independent, 26 Oct. 2007), while the Daily Telegraph highlighted Nathanson et al.'s argument ("Now women are told to keep off alcohol during pregnancy", Cockroft, Daily Telegraph, 26 Oct. 2007).

4.3   The Revision of the NICE Guidelines and Media Reaction

Even though the news coverage suggested a mistake by NICE because their guidelines were at odds with the DoH advice, it is worth remembering that these were only draft guidelines. In fact, one of the stakeholders who reviewed the draft noticed that there was a difference in tone with the DoH guidelines, and that this may lead to confusion with the public. It is not entirely clear though whether this stakeholder contribution was as a response to the media debate, though it is quite likely that the media reaction will have had an influence on the revision of the draft guidelines in any case. In the consultations comments table, the British Dietetic Association wrote.

We welcome a revision regarding alcohol, however there will be a confusion of messages to women as this does differ from the Department of Health advice. (NICE 2007b: 6)

NICE answered:

> Thank you. Following stakeholder consultation the recommendation has now been amended slightly so as to remain in line with the evidence but with the "safe" level of alcohol intake expressed in terms similar to that used by the DoH so as to avoid causing confusion. (ibid.)

This raises another issue – given that there are two bodies offering guidance on the same topic any change will inevitably result in a period of conflict between them.

As a result of the consultation, and possibly the negative press reaction to the draft, the final guidelines from NICE were changed when they were finally published in March 2008. In an accompanying document aimed at the public they wrote:

> If you are pregnant or planning to become pregnant, you should try to avoid alcohol completely in the first 3 months of pregnancy because there may be an increased risk of miscarriage. If you choose to drink while you are pregnant, you should drink no more than 1 or 2 UK units of alcohol once or twice a week. There is uncertainty about how much alcohol is safe to drink in pregnancy, but at this low level there is no evidence of any harm to the unborn baby. You should not get drunk or binge drink (drinking more than 7.5 UK units of alcohol on a single occasion) while you are pregnant because this can harm your unborn baby. (NICE 2008: 62)

Instead of being pleased that the NICE guidelines were changed to reflect the sorts of criticism they had attracted in the press, the media reaction to the published guidelines was one of puzzlement. In particular they seemed concerned that NICE would again change their advice, and expressed worry that the message for pregnant women had now become even more confused. For example, the headline in the Daily Mirror was "Your life: is booze a danger to unborn babies? (Can someone please decide)" (Jones et al., Mirror, 27 March 2008). It was also emphasised again that there is very little evidence that consuming low levels of alcohol does any harm.

The reaction of the Daily Mail can be considered angry; they suggested that the advice for pregnant women was now completely confusing.

> First the experts said you shouldn't drink. Then they said pregnant women could have a glass every so often. And after months of confusion, you might hope the health watchdog's latest advice on alcohol for expectant mothers would finally clear things up. But it seems nothing is that simple. (Hope, Daily Mail, 26 March 2008)

When NICE published their draft guidelines, they were clear that they were meant for consultation. Nevertheless, they were criticised for being in conflict with the DoH advice.

It seems slightly unfair that the subsequent publication of formal guidelines were then castigated for responding to this criticism and revising their draft guidelines in such a way as to ensure the advice the formally published was consistent with the perspective of the DoH. On the face of it, NICE appears to have listened to those who commented on the draft guidelines, taken note of the October press reports, and revised its advice so as to align with that issued by the DoH. NICE then faced a further accusation of engaging in a U-turn and, in so doing, not only causing confusing but also promoting the Nanny State. This reaction suggests, however, that the main issue for the papers was not the advice itself but a relatively minor lack of consistency between different sets of guidance on the same matter. "Experts disagreeing" sells papers, so we might think of this as something of a controversy manufactured by the press; however there seems to be also a more fundamental disagreement over the nature of public health risks that fuelled the controversy.

5 Discussion: Pitfalls in Advising on Low Risk but Widespread Behaviour

The press reaction to the updating of the advice demonstrates some of the pitfalls for public health bodies in issuing advice on behaviours that are widespread but low risk, particularly when there is no clear direction in which advice can unambiguously point. Whilst the scientific evidence for (or against) FAS is not entirely clear, there are strong moral arguments for preventing harm to the infant, as well as for thinking that pregnant women should be able to make their own decisions about consuming alcohol. Furthermore it's not entirely clear either what effect public health advice might have, or how the presentation of that advice might impact on the choices made by women. The ambiguousness of risks in science and the resulting difficulty in formulating clear policies, is a major part in the influential concept of post-normal science, where there can be several "legitimate perspectives" on the risks (Funtowicz and Ravetz 1993). From this perspective it is, then, unsurprising that in cases like FAS there might be no clear direction on which stance to take.

Despite the fact that debates about the use of alcohol in pregnancy have strong moral undertones (Armstrong 2003; Golden 2005) which, particularly in the US, seems to steer public debate towards the view that anything other than abstinence on the part of women who are pregnant or seeking to become pregnant is a morally questionable response. In this particular case, however, the initial reaction in the press was mostly one of puzzlement, or even hostility, towards the stronger formulations. There are other concerns that are very much present in contemporary public debates in the UK, which mean that discourse lean towards less prohibitive advice. This includes opposition to what is called the "Nanny State". In addition, in the relevant time period, the then Labour government was deeply unpopular. Whilst the government was not directly involved in drawing up the advice, comments made by the then public health minister, Dawn Primarolo, were very much in support of the stricter advice. These comments were given a good deal of coverage (e.g., Helm, Daily Telegraph, 9 Nov. 2007). One final factor that may have influenced the way the issue was covered in the national press was the fact, since its inception, NICE had been subject to fairly heavy criticism, perhaps most notably by the Daily Mail (Hawkes 2008). The BMJ's decision to present the issues in one of their Head to Head debates similarly shows how the conflict over the advice was seen as the main concern.

The one thing that seems to most clear from the press reaction to the advice published by NICE and the DoH was the presumption that scientific advice should be clear, unambiguous, and value neutral or free from anything that could be interpreted as moralising, which might be seen as slightly ironic given the press' often moralising stances when e.g. working class mothers or "laddettes" are concerned. In the absence of scientific clarity, the stance adopted by the UK media indicated that they would have liked to see a clear acknowledgement of the uncertainties about the state of our knowledge on what the safe limits are – this is why "experts disagreeing" in this case made such a good story, even if the actual disagreement was not, in the final analysis, all that great. Because of the failure to clearly acknowledge the uncertainty, the apparently conflicting advice given by the DoH and NICE – albeit in the form of draft guidelines – meant that the fact that we do not really know what the safe limits might be was more of an issue than might otherwise have been the case. This meant that it became hard to take the guidelines from either body seriously.

As a result Nice and, to a certain extent, the DoH were faced with a dilemma; they had to produce clear and unambiguous guidelines whilst, at the same time, being faithful to the scientific evidence. Unfortunately the scientific evidence simply does not point to any clear and unambiguous advice. To understand the dynamic between the press and the public health bodies in this story, we need to consider the different audiences they write for and pressures they come under. Rose's (1992) "prevention paradox", introduced above, is relevant here: If a large number of people slightly change their behaviour in such a way that what they do is a little less risky, then that saves more lives on average than if a small number of people substantially change a high risk behaviour. As a result, it is more reasonable for a public health body to try to persuade a great number of people to drink less than it is for them to, for example, persuade the very few number of parachute enthusiasts to stop throwing themselves out of aeroplanes. From the perspective of the individual member of the public however, it seems much less reasonable to concentrate on the relatively small risks almost all of us are exposed to at the expense of higher risks associated with a relatively small sub-population. From the perspective of public health bodies, stricter advice is rational because if successful it would save many children from FAS. For the individual, however, the change in risk will be barely perceptible. Thus, at one and the same time, it is perfectly rational for public health bodies to issue such advice, and it is perfectly rational for individuals to ignore it.

In this light, we can understand public health bodies as attempting to speak to the public at large, whilst the press can be seen as speaks to – or for – individual members of the public. Both perspectives are perfectly valid, but they are dependent on the ends one has in mind or the values that inform them. Different ends or values mean that one will adopt the point of view that best aligns with them.

6 Conclusions

Somewhat contrary to the expectations generated by lifestyle risk discourse research by Lupton (1993) – which presents such discourse as adopting a highly moralised stance towards risk-taking behaviour, especially when it puts others (in this case an unborn child) at risk – Connolly-Ahern and Broadway (2008) report that they found an emerging frame in the US newspaper coverage, one that was "in opposition to the prevailing abstinence frame". This study suggests that, at least in the context of this particular episode, this counter-frame was actually a dominant media reaction to public health advice on FAS in the UK. NICE was not criticised in the press for giving lax advice on FAS, but for being inconsistent with the DoH. It was the inconsistency that drew the most attention, even though the two sets of advice were not really that different.

Although some influential medical professionals such as Nathanson et al. (2007) argued that it is necessary to give strong simplified advice if the message being conveyed is to be clear, the press were more concerned about being given an accurate description of the risks. The UK media perceive the approach Nathanson advocates as patronising. Such concerns were later echoed by a bioethicist, who argues that this type of medical advice is ethically dubious because "it is not reasonable to replace more accurate information with less accurate merely because it is simpler to communicate" (Cavanagh 2009: 303). He also argues that this approach is also not in line with contemporary trends in medical ethics, which lean towards an increasing level of patient choice and autonomy.

For public health bodies this study may have certain implications that are worth considering when issuing public health advice. The immediately obvious one is that different bodies need to be aware that the advice offered should be consistent across the board – this is of course rather hard to do in practice and points instead towards the perils of having different public health bodies being tasked to issue advice on the same topics. Another point is that, when encountering the dilemma between giving advice that is clear and advice that is true to the scientific evidence, a frank admission of the fact that the evidence is not entirely clear may help to forestall accusations of inconsistency, although this may then in turn generate other accusations such as the "nanny state" issuing advice for no good reason.

However, a more subtle point I have tried to highlight here is the need to be aware of the various audiences and their interpretation of the issues – for example, the perception of risk will differ when considered from the perspective of an individual than when considered from the vantage point of a public health body. As I have suggested, this one of the reasons why the strengthening of the advice did not find much support in the newspapers. Awareness of this potential for conflict between the individual and public health perspectives of low but widespread risks will help anticipate possible reactions.

This study has been sociological and descriptive in that it attends to the discourse of various actors involved in the controversy. Given the focus of this volume on interrelating sociological and philosophical approaches (or, crudely put, to the descriptive and the normative), it is worth considering how any potential insights produced by this study might translated into normative guidance. How *should* we think about prevention paradox cases? With the prevention paradox being a central feature in explaining how various actors came to hold different views in this case, we can possibly turn to more abstract discussions on the bioethical impact of the paradox. Given how established the issue is in the public health literature, the philosophical examination of it is still only at the beginning stages.

John (2014) argues that the paradox provides us with a dilemma between "two prima facie plausible moral principles" – a public health intervention may reduce large risks of a few by introducing a much smaller harm (in this case in the form of foregone pleasures) on many? John proposes "ex-ante contractualism" as a moral theory that manages to capture most, but not all, of our moral intuitions regarding the paradox. By his own admission, John's treatment of the paradox remains messy and, as Thompson (2016) has shown, it can produce deeply counter-intuitive results in some circumstances.

This study has added a sociological analysis to a number of moral intuitions on a particular case as they appeared in the popular press. This, or so I hope, may be of use for the developing philosophical treatment of the paradox. In this case, the moral intuitions are perhaps more complex than the exemplars considered by John (2018) i.e. the concept of "minimal alcohol pricing" as a way of saving heavy drinkers at the expense of reducing enjoyment for "responsible" drinkers. The moral choice confronting pregnant women involves small potential risk to another (the fetus) rather than themselves. This demonstrates the value of this chapter in terms of the intentions in this volume; to build up a wider range of sociological case studies that demonstrate the variety of situations that the developing philosophical literature must address.

Acknowledgements The research was first carried out as part of the Winton Programme for the Public Understanding of Risk at the Statistical Laboratory of the University of Cambridge. Many thanks to David Spiegelhalter for inspiring this research, and to Ellie Lee and Jenny Bristow for comments on an earlier draft.

References

Abel, E. 1995. An Update on the Incidence of FAS: FAS Is Not an Equal Opportunity Birth Defect. *Neurotoxicology and Teratology* 17 (4): 437–443.
———. 1999. What Really Causes FAS? *Teratology* 59 (1): 4–6.
Allan, S. 2002. *Media, Risk and Science*. Buckingham: Open University Press.
American Academy of Pediatrics. 2000. Fetal Alcohol Syndrome and Alcohol-Related Neurodevelopmental Disorders. *Pediatrics* 106 (2): 358–361.

Armstrong, E. 2003. *Conceiving Risk, Bearing Responsibility: Fetal Alcohol Syndrome & the Diagnosis of Moral Disorder*. Baltimore: Johns Hopkins University Press.

Armstrong, E., and E. Abel. 2000. Fetal Alcohol Syndrome: The Origins of a Moral Panic. *Alcohol and Alcoholism* 325 (3): 276–282.

Boseley, S. 2007. Confusion Over Advice on Alcohol for Pregnant Women. *The Guardian*, October 11.

Boykoff, M., and J. Boykoff. 2004. Balance as Bias: Global Warming and the US Prestige Press. *Global Environmental Change* 14: 125–136.

Brashers, D.E. 2001. Communication and Uncertainty Management. *Journal of Communication* 51: 477–497.

Brown, J., S. Chapman, and D. Lupton. 1996. Infinitesimal Risk as a Public Health Crisis: News Media Coverage of a Doctor-Patient HIV Contact Tracing Investigation. *Social Science & Medicine* 43 (12): 1685–1695.

Cavanagh, C. 2009. "You Can't Handle the Truth": Medical Paternalism and Prenatal Alcohol Use. *Journal of Medical Ethics* 35: 300–303.

Clout, L. 2007. A Daily Tipple Is Safe for Women. *The Daily Telegraph*, October 11.

Cockroft, L. 2007. Now Women Are Told to Keep Off Alcohol During Pregnancy. *The Daily Telegraph*, October 26.

Connolly-Ahern, C., and S.C. Broadway. 2008. "To Booze or Not to Booze?" Newspaper Coverage of Fetal Alcohol Spectrum Disorders. *Science Communication* 29 (3): 362–385.

Curtis, P. 2007. Pregnant? Then Don't Touch Alcohol Is Latest Health Advice: Women Underestimate Dangers, Ministers Believe: Evidence in Support of Ban May Be Uncertain. *The Guardian*, May 25.

Daily Mail. 2007a. Why Our Nanny State Should Keep Mum. *The Daily Mail*, May 30.

———. 2007b. Here's Another Quasi-Scientific Solution to Make Mothers Feel Bad When Things Go Wrong.... *The Daily Mail*, October 16.

Day, K., B. Gough, and M. McFadden. 2004. "Warning! Alcohol Can Seriously Damage Your Feminine Health" A Discourse Analysis of Recent British Newspaper Coverage of Women and Drinking. *Feminist Media Studies* 4 (2): 165–183.

Department of Health. 2007. *The Pregnancy Book 2007*. http://webarchive.nationalarchives.gov.uk/+/http://www.dh.gov.uk/en/Publicationsandstatistics/Publications/PublicationsPolicyAndGuidance/DH_074920. Accessed 16 Oct 2017.

Friedman, S., K. Villamil, R. Suriano, and B. Egolf. 1996. Alar and Apples: Newspapers, Risk and Media Responsibility. *Public Understanding of Science* 5 (1): 1–20.

Funtowicz, S.O., and J.R. Ravetz. 1993. Science for the Post-Normal Age. *Futures* 25: 739–755.

Golden, J. 2005. *Message in a Bottle: The Making of Fetal Alcohol Disorder*. Cambridge, MA: Harvard University Press.

Harlap, S., and P.H. Shiono. 1980. Alcohol, Smoking, and Incidence of Spontaneous Abortions in the First and Second Trimester. *Lancet* 2 (8187): 173–176.

Hawkes, N. 2008. Why Is the Press So Nasty to NICE? *British Medical Journal* 337: a1906.

Helm, T. 2007. Don't Drink Alcohol, Pregnant Women Told. *The Daily Telegraph*, November 9.

Hope, J. 2008. The "Definitive" Guide to Pregnancy. But How Long Before the Experts Change Their Minds Again? *The Daily Mail*, March 26.

Independent. 2007. Pregnant Women "Could Drink". *The Independent*, October 26.

Institut National de Prévention et d'Éducation pour la Santé. 2007. 'Étiquetage: "Zéro Alcool Pendant la Grossesse"' Actualites: Lettre Bimestrielle sur les Effets de l'Alcool Oct. 35: 6. http://www.inpes.santepubliquefrance.fr/70000/dp/06/dp060911.pdf. Accessed 16 Oct 2017.

Inthorn, S., and T. Boyce. 2010. It's Disgusting How Much Salt You Eat! Television Discourses of Obesity, Health and Morality. *International Journal of Cultural Studies* 13 (1): 83–100.

Jensen, J.D. 2008. Scientific Uncertainty in News Coverage of Cancer Research: Effects of Hedging on Scientists' and Journalists' Credibility. *Human Communication Research* 34: 347–369.

John, S.D. 2014. Risk, Contractualism, and Rose's "Prevention Paradox". *Social Theory and Practice* 40 (1): 28–50.

John, S. 2018. Should We Punish Responsible Drinkers? Prevention, Paternalism and Categorization in Public Health. *Public Health Ethics* 11 (1): 35–44.

Jones, K.L., and D.W. Smith. 1973. Recognition of the Fetal Alcohol Syndrome in Early Pregnancy. *Lancet* 302 (7836): 999–1001.

Jones, C., C. Ward, and M. Bailey. 2008. Your Life: Is Booze a Danger to Unborn Babies? (Can Someone Please Decide?). *The Mirror*, March 27.

Kesmodel, U., K. Wisborg, S.F. Olsen, T.B. Henriksen, and N.J. Secher. 2002. Moderate Alcohol Intake in Pregnancy and the Risk of Spontaneous Abortion. *Alcohol and Alcoholism* 37 (1): 87–92.

Kitzinger, J. 1999. Researching Risk and the Media. *Health, Risk & Society* 1 (1): 55–70.

Kitzinger, J., and C. Williams. 2005. Forecasting Science Futures: Legitimising and Calming Fears in the Embryo Stem Cell Debate. *Social Science and Medicine* 61: 731–740.

Lee, E., R.M. Sutton, and B.L. Hartley. 2016. From Scientific Article to Press Release to Media Coverage: Advocating Alcohol Abstinence and Democratising Risk in a Story About Alcohol and Pregnancy. *Health, Risk & Society* 18 (5–6): 247–269.

Lowe, P., and E. Lee. 2010. Advocating Alcohol Abstinence to Pregnant Women: Some Observations About British Policy. *Health, Risk and Society* 12 (4): 301–311.

Lupton, D. 1993. Risk as a Moral Danger: The Social and Political Function of Risk Discourse in Public Health. *International Journal of Health Services* 23: 425–435.

Lupton, D., and S. Chapman. 1995. "A Healthy Lifestyle Might Be the Death of You": Discourses on Diet, Cholesterol Control and Heart Disease in the Press and Among the Lay Public. *Sociology of Health and Illness* 17: 477–494.

MacRae, F. 2007. Now the Advice Is Pregnant Women Can Have a Drink. *The Daily Mail*, October 11.

Nathanson, V., N. Jayesinghe, and G. Roycroft. 2007. Is It All Right for Women to Drink Small Amounts of Alcohol During Pregnancy? No. *British Medical Journal* 335 (7625): 857.

NICE. 2003. Antenatal Care (Replaced by CG 62). http://www.nice.org.uk/Guidance/CG6#implementation. Accessed 16 Oct 2017.

———. 2007a. Antenatal Care: Routine Care for the Healthy Pregnant Woman (Draft for Consultation). http://www.nice.org.uk/nicemedia/pdf/ANCpartial%20update2008Full%20Guideline.pdf. Accessed 1 Apr 2009.

———. 2007b. CG62 Consultation Comments Table. http://www.nice.org.uk/nicemedia/pdf/CG62ConsultationCommentsTable.pdf. Accessed 1 Apr 2009.

———. 2008. CG62 Antenatal Care: Understanding NICE Guidance. http://www.nice.org.uk/nicemedia/pdf/CG062PublicInfo.pdf. Accessed 16 Oct 2017.

O'Brian, P. 2007. Is It All Right for Women to Drink Small Amounts of Alcohol During Pregnancy? Yes. *British Medical Journal* 335 (7625): 856.

O'Leary, C., L. Heuzenroeder, E. Elliott, and C. Bower. 2007. A review of policies on alcohol use during pregnancy in Australia and other English-speaking countries, 2006. *Medical Journal of Australia* 186 (9): 466–471.

Riesch, H., and D.J. Spiegelhalter. 2011. "Careless Pork Costs Lives": Risk Stories from Science to Press Release to Media. *Health, Risk & Society* 13 (1): 47–64.

Rose, G. 1992. *The Strategy of Preventive Medicine*. Oxford: Oxford Scientific Publications.

Rose, D. 2007. Pregnant Women Told Glass of Wine a Day Is Fine – and Too Dangerous. *The Times*, October 11.

Rozin, P., and L. Singh. 1999. The Moralization of Cigarette Smoking in the United States. *Journal of Consumer Psychology* 8 (3): 321–337.

Ruhl, L. 1999. Liberal Governance and Prenatal Care: Risk and Regulation in Pregnancy. *Economy and Society* 28 (1): 95–117.

Shaikh, T. 2007. No Alcohol At All During Pregnancy, Doctors Say. *The Independent*, May 25.

Stocking, S.H. 1999. How Journalists Deal with Scientific Uncertainty. In *Communicating Uncertainty: Media Coverage of New and Uncertain Science*, ed. S. Friedman, S. Dunwoody, and C. Rogers. Mahwah: Lawrence Erlbaum Associates.

Stocking, S.H., and L.W. Holstein. 2006. Manufacturing Doubt: Journalists' Roles and the Construction of Ignorance in a Scientific Controversy. *Public Understanding of Science* 18: 23–42.

Thom, B. 1994. Women and Alcohol: The Emergence of a Risk Group. In *Gender, Drink and Drugs*, ed. M. McDonald. Oxford: Berg.

Thompson, C. 2016. Rose's Prevention Paradox. *Journal of Applied Philosophy*. https://doi.org/10.1111/japp.12177.

Wahlberg, A., and L. Sjöberg. 2000. Risk Perception and the Media. *Journal of Risk Research* 3 (1): 31–50.

Washer, P. 2004. Representations of SARS in the British Newspapers. *Social Science and Medicine* 59: 2561–2571.

Weingart, P., A. Engels, and P. Pansegrau. 2000. Risks of Communication: Discourses on Climate Change in Science, Politics, and the Mass Media. *Public Understanding of Science* 9: 261–283.

The Ethical Framework for the Use of E-Cigarettes

Nancy Tamimi

1 Background on E-Cigarettes

E-cigarettes are battery-powered devices; they were invented by a Chinese pharmacist and were first marketed as an alternative to regular smoking and an aid to stop smoking. They received their first international patent in 2007 (Caponnetto et al. 2012). They mainly vaporize a mixture of water, propylene glycol or glycerin, flavourings, with or without nicotine, and are activated when the user inhales through the mouthpiece of the device. No tobacco or combustion is necessary for their operation. Different brands, designs and generations became available in the market with different prices, sizes, colours, weights, accessories, flavours and with variable levels of nicotine. Since 1985, several products have been introduced to the market as non-therapeutic devices, but a substitute to smoking. All the products were mainly made by tobacco companies and were not accepted by regulators and public health professionals (Orleans and Slade 1993: 18). The Swedish snus is another smoking replacement that has been met with vigorous opposition from tobacco control activists and regulators, and its use is still prohibited in most European countries including the UK (Gartner et al. 2007; Hajek 2015). To the contrary, the popularity of e-cigarettes has increased nationally and internationally (Adkison et al. 2013; West et al. 2014). In 2015, it was estimated that the number of e-cigarettes current users in Great Britain reached 2.8 million, mostly made of current and ex-smokers (ASH 2016: 1).

A growing body of international and national studies has investigated the reasons for e-cigarettes use (e.g. Farsalinos et al. 2014; Richardson et al. 2014; Biener and Hargraves 2014; Hummel et al. 2015; Pepper et al. 2015; ASH 2016). In general, it was noted that across different populations, the top endorsed reasons for using e-cigarettes were: to help stop smoking and harm reduction (McNeill et al. 2015:

N. Tamimi (✉)
Department of Global Health & Social Medicine, King's College London, London, UK
e-mail: nancy.k.tamimi@kcl.ac.uk

© Springer International Publishing AG, part of Springer Nature 2018   127
H. Riesch et al. (eds.), *Philosophies and Sociologies of Bioethics*,
https://doi.org/10.1007/978-3-319-92738-1_8

53). Their possible efficacy, it was argued, could be attributed to e-cigarettes' nicotine delivery feature, coincidently addressing the chemical aspect of nicotine addiction and imitating the habit of smoking cigarettes to maintain the smoking behaviour (Farsalinos et al. 2013). Although existing research does not provide a definitive conclusion about e-cigarettes' safety in absolute terms, there is an indication that they are much safer than tobacco cigarettes, and comparable in toxicity to approved Nicotine Replacement Therapy (NRT) (NICE 2013a: 11; Goniewicz et al. 2014). Public Health England (PHE) announced that e-cigarettes could be 95% less harmful than cigarettes (PHE 2015: 5).

E-cigarettes are becoming socially attractive and part of a developing trend; they are publicised as 'lifestyle products', endorsed heavily by celebrity involvement and promotions on the internet and different forms of social media (Bauld et al. 2014: 10). In the UK, e-cigarettes used to be marketed as a less harmful substitute for traditional tobacco cigarettes and were regulated under the General Products Safety Directive (de Andrade and Hastings 2013). From May 2016, two types of e-cigarettes became available, consumer products and licensed medicine.

E-cigarettes, however, have created a 'moral quandary' (The Lancet 2013: 914). This is augmented because of the similarity in the way e-cigarettes are consumed compared to smoking cigarettes; the use of promotional images and messages, to advertise for e-cigarettes, similar to those used in the past to promote traditional cigarettes as well as the growing involvement of the tobacco industry in manufacturing and marketing e-cigarettes (de Andrade et al. 2013; de Andrade and Hastings 2013). The literature has discussed the 'bioethicists' uncritical acceptance of scientific futures' (Hedgecoe 2010: 165). Nevertheless, bioethics, as a moral ethical theory, remains a key at identifying and exploring the ethical issues raised by new technologies (Callahan 1999). I will discuss the ethical issues raised by nicotine addiction, anti-smoking regulations and the use of e-cigarettes with a particular link to the principlism bioethical framework. Principlism was first proposed in 1977 by Beauchamp and Childress and consists of four moral principles: autonomy, beneficence, non-maleficence and justice (Beauchamp and Childress 2009). Since its introduction, principlism has been widely used and discussed, despite its limitations and criticisms, both in practice and in the academic literature (Samuel and Brosnan 2011; Page 2012). Beauchamp and Childress claim that their four principles offer a simple ethical framework which enables identifying, analysing, examining, and addressing the increasingly complex ethical concerns that have emerged in bioethics, especially in health care situations. Some scholars consider principlism to be an important tool for legislators and public policy makers, who provide regulations for individuals, groups and institutions with different cultural backgrounds, worldviews and agendas (McCarthy 2003: 69; Jarvie and Malone 2008). Principlism was described as an 'abstract, universal, comprehensive, detached, impartial, objective, neat, exact, clear, coherent system of ethical decision making' (McCarthy 2003: 69). However, in morally difficult situations where there is a conflict between principles, each principle is ought to be weighed according to the particular context in which it is applied, without favouring one principle over another (Beauchamp and Childress 2001: 398).

There has been a lack of social sciences' contribution to bioethics in previous decades. Some social scientists have criticised bioethics as an abstract and individualistic concept and called for a better way, than principlism, in which a society can make ethical decisions in the face of biomedical dilemmas (Evans 2000). However, other scholars called for a focus on the common concerns between bioethics and social sciences (Turner 2009), and others emphasised the changes in the relations between the empirical and normative perspectives on bioethics (Borry et al. 2005). In spite of its criticisms, the value and applicability of principlism in addressing different areas of ethical concerns have dominated decision making in the West, and hence my choice to apply the approach to e-cigarettes use at times where the moral quandary about its use is ostensible.

Before discussing the literature of bioethics in relation to tobacco and e-cigarettes use, it is useful to explain briefly the current views on nicotine addiction and tobacco control regulations in the UK.

2 The Development of Nicotine Addiction and Tobacco Control Regulations in the UK

Nicotine is derived from the tobacco plant. It is suggested that tobacco smoking in Europe began around 1560 (Doll 2004). Following the invention of the cigarette-making machine in 1881, cigarette smoking became the most preferred and widespread way to consume tobacco in the West (Berridge 1999). During the 1960s, leading health organisations, such as the Royal College of Physicians (RCP), the advisory committee to the US Surgeon General on Smoking and Health and the World Health Organization (WHO), viewed smoking as a sign of "habituation", where social factors and personal characteristics influence the habit rather than biological addiction. In spite of some attempts by scientists at the time to draw attention to the role of nicotine in addiction, it was not until the late 1970s that such leading health organisations started to view smoking as a form of addiction. Only in 1988 cigarettes were declared to be addictive and nicotine as the main cause of addiction due to more proven psychological and physiological effects of smoking (Parascandola 2005; Keane 2013: 190). In 1964, the WHO replaced the term 'addiction' with the term 'dependence'. The 'dependence syndrome' was defined as:

> a cluster of physiological, behavioural, and cognitive phenomena in which the use of a substance or a class of substances takes on a much higher priority for a given individual than other behaviours that once had greater value. A central descriptive characteristic of the dependence syndrome is the desire (often strong, sometimes overpowering) to take the psychoactive drugs (which may or not have been medically prescribed), alcohol, or tobacco (WHO 2014).

Over the years both terms 'nicotine dependence' and 'nicotine addiction' became widely accepted when describing tobacco use with no clear distinction between the two terms (Keane 2013). Moreover, the 1988 Surgeon General's Report on *The*

Health Consequences of Smoking provided evidence to underpin the notion that nicotine is addictive like heroin and cocaine (U.S. DHHS 1988: 281). Indeed, the WHO states that '[n]icotine meets the established criteria for a drug that produces addiction, specifically, dependence and withdrawal' (WHO 2010: 141). A comparison with other addictive drugs found that it is more common to develop a dependence on nicotine than that on cocaine, heroin, or alcohol. While 'rates and patterns of relapse' are similar for all of them, 'tobacco was associated with equal or greater levels of difficulty in quitting and urge to use' (Henningfield and Benowitz, 2004: 139). It was found that deprivation from tobacco use produces withdrawal symptoms that are 'intermediate between that of opioids and cocaine'. This may include: irritability; depressed mood; anxiety and sleep disturbances, and although they do not pose a risk to life, they can be 'occupationally and socially debilitating' (Henningfield and Benowitz 2004: 139; Glover 2006).

Although some scholars challenged the evidence of the addictiveness of nicotine and called for re-examining the 'nicotine addiction thesis' (Dar and Frenk 2004), in the UK, nicotine is generally believed to be a highly addictive substance but not a carcinogen (RCP 2007: xi; 2016: 5; PHE 2014: 7). Further, there has been no consensus, among scholars, on the definition and models to explain addiction (West and Brown 2013). Nevertheless, the biomedical model became widely accepted, especially after neuroscience research had provided evidence that addiction has a neurobiological basis. Studies showed that the major psychoactive drugs of dependence act on neurotransmitter systems in the brain (Wayne et al. 2003; Dackis and O'Brien 2005; Buchman et al. 2010). It is proposed that the addictive nature of nicotine is attributed to the activation of the reward pathways in the brain, resulting in increased secretion of the neurotransmitter dopamine, which regulates the feelings of pleasure experienced by many smokers. When smoking a cigarette, the nicotine reaches the brain within 10 s of inhalation. However, this effect of pleasure disappears quickly, and consequently causes smokers to continue smoking to maintain the reward effect and prevent withdrawal symptoms, and thus maintaining tobacco addiction (Markou 2008). This has led to increasingly characterising nicotine addiction as a 'disease of the brain' (Wise 2000; Glover 2006). Further, although the habit of smoking is typically initiated voluntarily. The continuation of the habit has been attributed to self-control impairment. This has been linked to changes in the brain. Brain imaging studies of people with addiction showed

> physical changes in areas of the brain that are critical to judgment, decision making, learning and memory, and behavior control, which scientists believe may help explain the compulsive and destructive behaviors of addiction (National Institutes of Health U.S. 2007: 7).

These findings are important when discussing the autonomy of addicts in bioethical contexts as will become clearer later in the chapter.

During the 1970s, Professor Michael Russell explored the role of nicotine in smoking; he concluded that "[p]eople smoke for nicotine but they die from the tar" (Russell 1976: 1431). Russell's findings that 'nicotine addiction is not harmful' (1991: 654) led him to promote the use of pharmacological 'clean' nicotine as a self -administration tool to replace the 'dirty' tobacco and fight smoking related diseases (Russell 1976: 1431). In 1998, the Department of Health published the White Paper, *'Smoking Kills'* which led to the implementation of a variety of tobacco-control policies, including a ban on tobacco advertisements, increasing the price of tobacco, a ban on smoking in workplaces and enclosed public places. It also led to the formation of the National Health Service (NHS) Stop Smoking Services (SSS) to support smokers who seek their help to quit smoking. The main strategy was for Stop Smoking Advisors to provide behavioural support, a short-term use of Nicotine Replacement Therapy, or other licenced treatments, with an emphasis on abstinence-from smoking-only strategy (DOH 1998: chapter 4). While cigarettes were described as 'very 'dirty' delivery systems for nicotine' (ASH 2004), 'pure nicotine products', 'although addictive' were viewed to be 'considerably less harmful' and were promoted to help people quit smoking (ASH 2013: 1, 2014). PHE (2014: 7) accentuated the separation between nicotine and smoking by confirming that 'nicotine is not a significant health hazard. Nicotine does not cause serious adverse health effects such as acute cardiac events, coronary heart disease or cerebrovascular disease, and is not carcinogenic'. In the same vein, sociological scholars described the moral dichotomy of nicotine where 'good' '(health promoting)' nicotine represents the use of the pharmaceutical form of nicotine, and 'bad' '(health destroying)' nicotine represents the recreational use in the form of cigarette smoking (Bell and Keane 2012: 245; Keane 2013: 190).

In 2011, the new tobacco control strategy for England incorporated '[n]ew approaches to help tobacco users who cannot quit to instead use safer sources of nicotine' (DOH 2011: 36). The UK Medicines and Healthcare products Regulatory Agency (MHRA) approved an extension to the indication of Nicotine Replacement Therapy to include 'harm reduction' to be used 'as a complete or partial substitute for smoking', both for those making an attempt to quit and those not currently intending to make a quit attempt but wish to reduce smoking, 'without any restriction on its duration of use' (MHRA 2010). Later a ban on smoking in cars was introduced and warning labelling progressed to plain cigarettes packaging. The next section explores the resulting ethical issues raised by contemporary views on nicotine addiction and tobacco-control policies.

3 The Ethical Issues of Views on Nicotine Addiction and Tobacco-Control Policies

Some ethical issues were raised in relation to, first, classifying nicotine use as a form of addiction. The concept of addiction, it was argued, was invented in the late 18th and early nineteenth century as part of a fundamental social structural transformation, which increased the focus on individual's control over their own 'compulsive' or 'deviant' behaviours (Levine 1985: 53). During the early twentieth century, people addicted to drugs were viewed as 'morally flawed and lacking in willpower' rather than having a health problem; thus, there was an emphasis on disciplinary measures rather than prevention and therapy (National Institute on Drug Abuse 2008: 1). It was argued that the term 'dependence' has replaced 'addiction' to enforce the medical authority and to overcome the moralistic and judgemental associations with the word addiction (Keane 2013: 191).

Second, the medical explanation of addiction. Some scholars expressed their concerns that placing addiction within the biomedical model and 'the biological' explanation is 'displacing the ethical', by *crowding out* or *diminishing* the 'moral responsibility for one's conduct' (The President's Council on Bioethics 2003). The brain disease model, it was argued, 'further implies simplistic categorical ideas of responsibility', where addicted individuals have no control over their substance use. Consequently, this may, unintentionally, influence individual's 'sense of identity, responsibility, notions of agency and autonomy, illness, and treatment preference' (Buchman et al. 2010: 2). The latter scholars described how the purely reductive neurobiological explanations of addiction legitimize addiction as a medical condition, ignoring the wider social context influences such as psychological, social, political, and other factors. The neurobiological advances, it is argued, has '*re-fashioned*' addiction as a medical condition that needs medical intervention, and enforced biomedicalization of addiction (Campbell 2012: 15–16). Also, although the neuroscience explanations for addiction has the potential to improve treatments for drug dependence, some authors believed that stigma may increase for addicted people because biogenetic causal theories lead to increase perceptions of 'dangerousness, unpredictability and desire for social distance' (Read et al. 2006).

Third, tobacco control policies. The ethical foundation for the smoking ban policies and the moral drive for the goals of public health, and what constitutes a final victory in tobacco control policies, were questioned (Shickle 2009: 20; Thomas and Gostin 2013). Warner (2013: 22) wondered if the end goal for such policies would be the end of all smoking, or the end of all forms of tobacco use, or a nicotine-free population. He questioned if achieving a more modest tobacco harm reduction objective would complete the endgame. Pope (2000: 424) also asked: 'where does one draw the line between the power of the state to protect public health and the right of the people to make lifestyle choice'. Further questions were raised with regards to the morality of justifying the override of individual freedom for the sake of promoting the health of the population; thus treating individuals 'as less than moral equals' and denies them 'the right to choose their own ends of action' (Buchanan 2008: 16). Stopping smoking, it was argued, should be the responsibility of smokers themselves (Verweij 2009: 185).

According to Beauchamp and Childress principle, respecting smokers' autonomy involves 'acknowledging the value and decision making rights of persons and enabling them to act autonomously' (Beauchamp and Childress 2009: 103). Hence, 'being free of coercion (voluntariness)' is a necessary requirement for autonomy (Pope 2000: 460). Nevertheless, Beauchamp and Childress (2009: 105) explained that the principle of 'respect for autonomy' does not apply to individuals 'who cannot act in a sufficiently autonomous manner (and who cannot be rendered autonomous)', such as drug dependent patients. There is a considerable disagreement on the extent to which addicted smokers have control over their smoking behavior.

However, the biological explanation of addiction, the nature of addiction and the psychological suffering that smokers face when they try to quit, suggest 'that addiction may undermine autonomy' (Scragg et al. 2008). So the decision to continue to smoke is considered by some scholars to be non-autonomous (Pope 2000: 468). Hence, paternalistic policies, such as a complete ban on tobacco products or restricting smoking in some places, could be justified if nicotine addiction is viewed as a non-voluntary action, especially with the knowledge of their harmful effect (Jarvie and Malone 2008; Verweij 2009).

Bean (2014: 4) explained how anti-smoking interventions, among other public health interventions, are largely based on the harm principle; following John Stuart Mill's view: '...the only purpose for which power can be rightfully exercised over any member of a civilized community, against his will is to prevent harm to others.' (Mill 1859). Hence, the use of coercion to restrict the actions of individuals is justified 'to prevent unacceptable harm to entities worthy of protection' (Jensen 2002: 40). For example, in their discussion of the issue of restricting parental smoking in private domains to protect children from the harms of second-hand smoke, Jarvie and Malone emphasised that it is ethically justifiable to temporarily and intermittently restrict the autonomy of an autonomous adult smoking near children. They argued that such restriction represents an act of beneficence (a duty to prevent or remove evil or harm and to promote good to protect the child) and nonmaleficence (obligation to not inflict evil or harm towards children). However, the authors argued that forcing adults to stop smoking as a way to protect children from second-hand smoke exposure 'would constitute an unjustifiably paternalistic approach' (Jarvie and Malone 2008: 2144).

Moreover, it has been argued that in their assessment, tobacco control policy makers compare the short-term pleasure of smoking to the longer terms poor quality and quantity of life which they seek to improve (Shickle 2009: 9). However, Lieber and Millum (2013) argued that health and economic benefits are not the only scopes of welfare to smokers. They pointed at the freedom of choice that an individual should be guaranteed to experience any kind of goods associated with a particular action. Smoking, the authors argued,

> is associated with many subjectively valued goods, including pleasurable physiological effects (such as a smoking high, stress reduction, and appetite suppression) and non-physiological effects (such as projecting a desired image and the sociological benefits of being part of a group unified by a characteristic behavior) (Lieber and Millum 2013: 26).

It is these 'subjectively valued goods' that e-cigarettes were able to preserve and it is in this context that the ethical debate about the use of e-cigarettes arises. I will be reflecting on the findings of a sociological study which applied semi-structured interviews to explore the meanings and perceptions of e-cigarettes among15 e-cigarettes users and 13 Stop Smoking Advisors in South East England. Sociological research is needed to help bring to light some of the ethical issues surrounding the use of e-cigarettes. Many social sciences scholars criticised the principle-centred bioethics approach, for de-contextualising the process of ethical decision-making (Samuel 2014: 183). They highlighted the need for sociological research to increase the contribution of bioethics to public debate and policy on emergent technological innovations (Fox 1976; Hoffmaster 1992; Haimes 2002; Hedgecoe 2004; Samuel and Brosnan 2011; Williams and Wainwright 2010). Sociological studies can unravel the role of social and cultural factors, and hence provide a big picture of the way the ethical reasoning is happening and shaping the new and emerging technologies (Hedgecoe 2004). Haimes (2002) described the theories of prominent social scientists, like Foucault and Lash, and demonstrated how ethics are embedded within various aspects of the social world such as family, gender, health, institutions, community, economy, sexualities, religion, law and politics. Social Sciences, it has been argued,

> see legal and ethical issues as primarily social issues and, because of this encompassing perspective, can contribute not only to the understanding of ethical issues but also to the understanding of the social processes through which those issues become constituted as ethical concerns (Haimes 2002: 91).

Although this study has not been set out to explicitly investigate the ethical issues of e-cigarettes per se, but it nonetheless raised some ethical questions with regards to e-cigarettes. Klieman (1999: 71) highlighted that 'empirical research can provide knowledge about local worlds of experience', and affirmed the importance of these 'local moral processes' for bioethics. Interviews can elicit the diverse range of definitions and meanings that underpin both the ethical dilemma that surrounds e-cigarettes use and the answers offered by different actors (Haimes 2002). Hence, this sociological empirical work has the potential to enhance our understanding of some of the ethical issues that surround e-cigarettes. The next section will examine how ethical principles might apply in balancing the responsibilities of society and individual adults in relation to e-cigarettes. Their use will be explored through discussing the ethical issues of Harm Reduction (HR).

4   The Ethical Issues of Harm Reduction (HR) and E-Cigarettes

Harm reduction is defined as:

> [A] social policy which prioritizes the aim of decreasing the negative effects of drug use as an alternative drug policy to abstentionism, which prioritizes the aim of decreasing the prevalence or incidence of drug use (Newcombe 1992: 1)

Newcombe (2015) perceives HR as a strategy with a hierarchy of goals. For example, to reduce the transmission of HIV infection among injecting drug users, HR strategies may include stop sharing injection equipment, reduce the quantity of consumed drugs and abstinence. Therefore targeting consumption here is viewed as one goal for HR approach. Aceijas (2012: 22) explained how HR conforms to the bioethics' principles,

> The well-established principles of HR—pragmatism, goal prioritization, humanism, focus on harms and risks, no focus on abstinence and maximization of the range of available intervention options—have been portrayed as weak and evidence of a defeatist attitude towards drugs. They, however obviously match the main principles of bioethics in health care.

Scholars differentiated between the ethical foundation of abstentionism and HR which

> has its main roots in the scientific public health model, with deeper roots in humanitarian-
> ism and libertarianism. It therefore contrasts with abstentionism, which is rooted more in
> the punitive law enforcement model, and in medical and religious paternalism (Newcombe
> 1992: 1).

In the field of smoking, The National Institute for Health and Clinical Excellence (NICE) guidelines explained that smokers can reduce the harm of smoking for themselves and for the people around them, by stopping smoking altogether, cutting down prior to quitting, smoking less or abstaining from smoking temporarily (NICE 2012: 4). NICE also pointed at the advantage of using licensed nicotine-containing products, either temporary or for the long-term, because, they stated, '[u]sing these products can make it easier for people to cut down before stopping, reduce their smoking or abstain"(NICE 2013a, 6.8; b). HR advocates have questioned the different positions that public health programmes take when applying HR. While it is acceptable to apply HR for injecting drug users, it is rejected when dealing with smokers, instead, tobacco control policies, as clarified earlier, insist on the abstinence-only strategy.

It is argued that smoking reduction, which should result in reducing risk, rather than smoking elimination, is a 'moral imperative' (Fairchild et al. 2014: 295). In the literature, the cutting down approach was discussed and had raised the ethical concern of maintaining the addiction and the harm of tobacco addiction in society (Roe 2005; Cahn and Siegel 2011; RCP 2007). This concern was highlighted among the advisors in this study as the following example reveals,

> I can see that having the reducing over a period of time might help, however, what we know
> about the receptors in the brain, the more you are feeding it the nicotine the more it wants
> it. I think therefore I am not certain whether that slowing, … stopping over a longer period
> of time, ultimately, is that going to help them or not (H5, Advisor)

However, it was acknowledged that maintaining nicotine addiction does not connote maintaining the harm,

> I think there is a misunderstanding there, the harm that nicotine does, it is not dangerous
> chemical as such. Ok, it can cause perhaps stress to the cardiovascular system, but in the
> main, it's all the other chemicals in the cigarettes which cause the damage and if it did cause
> harm then it would not be licensed as a long-term use as a medication… I don't think it
> maintains harm. I think it does maintain addiction, yes, you are still depending on maintain-
> ing that product I suppose or maintaining that substance. But I don't think it's harmful it's
> not being shown to cause cancer or anything like that (H6, Advisor)

The long-term use of nicotine products was a controversial issue. Although for some advisors, quitting smoking was the main aim, for others, it is not only the smoking cessation that they were aiming to achieve, it is the elimination of nicotine addiction. E-cigarettes are used by many smokers to stop smoking cigarettes; however, the pattern of their usage differs. In my study, some used e-cigarettes with a view to quitting their use at some point in the future, others used them as a substitute to tobacco smoking without the intention to quit them. However, the Stop Smoking

Advisors brought e-cigarettes into the world of medicinal nicotine, they fitted e-cigarettes within the Nicotine Replacement Therapy framework.

> I think they [e-cigarettes] are effective maybe as a combination of NRT or just slightly less....Sometimes they will come to the [prescription] list. I think they will be another piece in our armoury (S2, Advisor).

So, generally, the advisors favoured a short-term use of e-cigarettes with the end goal of ending both e-cigarettes use and nicotine addiction as the following examples reveal,

> If someone comes to the clinic with me and they're smoking an e-cigarette I always discuss with them what kind of level the nicotine is and then we can look at weaning them off the nicotine low or 0 nicotine and products that still use the vape but without the nicotine in it, you know, it's the ultimate goal (S3, Advisor).

> No not long term, I would say after like 10 weeks, I want you to cut down, I want you to lower the dose.... because it is about them taking control of the behaviour. This is the key point that our public health, our team.... wants that person to [achieve] empowerment and control and confidence and esteem... all what we do just substitute. I think that is the focus, we might say to them ok use e-cigarette first 4 weeks, monitor on a weekly basis,... week six we are going to cut down so over 12 weeks give up smoking and not using anything at all' (H3, Advisor)

This view is problematic, because as pure nicotine is believed, by main UK health organisations, to be a relatively safe product, there is no obvious moral basis to lead smokers to quit nicotine.

It is suggested that a 'cultural anxieties' have developed over e-cigarettes in the West; that

> the war on smoking became cultural, with disapproval and ostracism of anyone who lit up; and that, although e- cigarettes have removed the war's scientific basis, our cultural revulsion persists. Therefore, so does our prohibition and condemnation (Saletan 2009).

E-cigarettes, it has been suggested, have created 'new moral panic' (Marcotte 2014). The panic is no longer over tobacco links to increased morbidity and mortality in societies, but over nicotine addiction by itself as the following advisor said:

> It's a chemical that they don't need, it's not going to benefit them so that's that the danger it may open up a door for an addiction they did not have, to begin with. ... It's a double edge sword (H1, Advisor)

Nicotine addiction is viewed as a bad habit because addiction per se is considered to be malign. It has been argued, that the resistance to accepting e-cigarettes and the public performance of the habit is shifting the stigma associated with tobacco cigarette use to e-cigarettes (Shickle 2009: 20) as the following user explained,

> I think generally there is less of a stigma around electronic cigarettes, and that certainly wasn't in the beginning anyway, when it was still seen as a novelty and people.. started using them as a way of quitting and it was supposed to be a healthy thing, and I think there was no stigma. Now that it's become kind of lifestyle thing in itself, I think the stigma has returned a bit. And this kind of links up to... just generally a stigma and the people judging

other people's lifestyle choices. In the end… it is a lifestyle choice and I think the stigma that smokers had has kind of carried on to the e-cigarette users in the sense that the stigma surrounding the addiction itself. People see you as weak because you give in to the addiction (12L, e-cigarette user).

Scholars argued that the panic over e-cigarettes in particular, apart from other nicotine delivery products, such as patches and gums, stems from opposition to the visibility of any smoking like behaviour (Bell and Keane 2012: 246; de Andrade et al. 2013; Marcotte 2014) as the following extract confirms,

The other aspect to it is for the e-cigarettes that look like a cigarette. It isn't doing anything to stop the hand to mouth action; it's perpetuating that habit; it looks the same as a cigarette for babies, children and young people from a distance; it still looks the same type of activity doesn't it? … I am not quite so certain about that as a method to stop smoking (H5, Advisor)

However, most my interviewees (from both groups) considered vaping (the use of e-cigarettes) to be more socially acceptable than smoking. For example, for one user (2B) 'social life is better' because he no longer needed to leave the room to have a 'fag'. Another user emphasised the encouragements he received from his family for using e-cigarette and how he made his daughter happy for using e-cigarette rather than cigarettes (13M). One advisor predicted that once e-cigarettes become regulated, it is going to become 'more cultural norm' (S3). So how the different regulations on e-cigarettes are viewed within the principlism framework.

5 E-Cigarettes Policies and Principlism

According to the World Health Organisation Framework convention on Tobacco control, policies that promote e-cigarettes as an alternative to smoking and as an aid to quit smoking simultaneously have the following implication. First, they imply that e-cigarettes can be used as the last resort to help quit smoking; second, that there is no need to quit nicotine addiction, just smoking; and lastly, that e-cigarettes can be used where smoking is not possible. The convention asserts that 'some of these messages are difficult to harmonize with the core tobacco-control message and others are simply incompatible' (WHO FCTC 2014: 7).

In view of the principlism framework, there are some ethical issues to consider from banning the use of e-cigarettes. First, in terms of autonomy, 'prohibiting e-cig[arette] infringes on smokers' autonomy to use a less harmful nicotine product while inconsistently allowing individuals to begin and continue smoking cigarettes' (Hall et al. 2015: 1). This interferes with the rational adults' actions or behaviours as the following extract from one e-cigarette user shows,

My opinion and that's with my fellow people who smoke/vapes, is we're quite angry about the idea that it's going to be banned and that legislation is going to be brought in, and we're really quite angry, because those of us who have managed to keep off tobacco, know that we're still addicted, but we found this to be much more pleasant in order to deal with our addiction. So if they start banning it, it's just going to drive it underground, I guess (14N, e-cigarette user)

Allowing e-cigarettes use as a substitution to smoking, Hall and co-authors (2015) believe, respects the autonomy of users, who take their own decision and keep their right to use the addictive substance based on their subjective preference, and the belief that harm is less for themselves and for others. Using e- cigarettes as an aid to stop smoking, outside the health care jurisdiction, is the ultimate instance of autonomy for individuals in a democratic liberal society. So, while some smokers can decide to use e-cigarettes as a tool to help them quit smoking, others, who do not want or not ready to give up smoking, can substitute to a safer alternative. Bean (2014) states that 'an individual's ability to self-determine may be affected if we curtail their ability to implement potential harm reduction measures such as choosing e-cigarettes over traditional cigarettes'.

Nevertheless, as discussed before, there is still a disagreement onto whether the choice to continue the habit of consuming nicotine is autonomous. Testimonials from my research highlighted the issue of the formation of involuntary addiction to e-cigarettes

> It's going to be as big a struggle as giving up normal cigarettes. ..But then, I suppose, in a sense..., they help you stop smoking, I can see – I think for me they mimic smoking almost perfectly, so the addiction is almost as bad to these things as it is to cigarettes. So I don't know if I do manage to quit these and I will need willpower of exactly the same scale (12L, user)

> Cure the nicotine addiction no way .....I don't think the tool itself is good enough to use it like NRT. They have to do a lot of work and regulation to be able to say this is a medication to cure it. I think it's just another substituting and the smoking habit can be changed yes but the nicotine dependency itself I doubt that (H4, Advisor)

In terms of non-maleficence, Hall et al. (2015: 1) believe that e-cigarettes ban 'perpetuates harm by preventing addicted smokers from using a less harmful nicotine product'. While the action of substituting tobacco with e-cigarettes reduces harm to others, the long term harm that e-cigarettes use may cause to others is still unclear (Hajek et al. 2014). The evidence so far suggests that e- cigarettes' vapour is much less harmful than second-hand smoking (Czogala et al. 2014; NCSCT 2016) or negligible (Burstyn 2014). Beside the second-hand vapour risk, other risks to others have been related to e-cigarettes and contested, these are, normalising smoking, gateway risks and some safety concerns related to the quality of the products. However, those risks are not proved in absolute terms (de Andrade and Hastings 2013; NCSCT 2016; Bauld 2016), and it was suggested that regulations can improve the quality issues. Hence, Bean (2014) asked if it is

> [F]air to restrict access to a product that may be helpful for a minority of the population (i.e., smokers) based on concerns for the safety of others (e.g., second-hand smoke, smoking re-normalization), particularly based on limited evidence.

In terms of the principle of positive beneficence, smokers who substitute tobacco with e-cigarettes, take positive steps to contribute to the welfare of others by eliminating the harms from second-hand smoke. As of utility, e-cigarettes provide help to smokers who tried and failed to stop smoking over the years and do not feel prepared to give up their addiction.). E-cigarettes ban, Hall et al. (2015: 3) argued, is against the principle of equity, where people who want to use e-cigarettes to help them quit smoking are deprived of this opportunity. E-cigarettes may be a tool that produces the greatest good for the maximum number, where values such as pleasure, happiness and preference satisfaction are maximised. Similar to the success of the opiate substitution therapeutic options, allowing the use of e-cigarettes will reach populations of addicts, who other ways of stopping smoking have failed them. Therefore contribute to their welfare and the welfare of the wider society. The following user shed light on what e-cigarettes bring to smokers,

> I guess the electronic cigarette helps you to stop smoking because... stopping smoking is difficult for a couple of reasons, ...one of them is about you're addicted to nicotine. The electronic cigarette will do that. It will cover that problem quite well and I think it would. The other thing about stopping smoking is about the satisfaction of the activity of smoking....., other aids, to my mind, my experience, don't give you that. You chew a piece of gum, which you're not even allowed to chew, you've got to keep in your cheek and it produces all these vile stuff and you've got to try not to swallow. It's horrid. It's disgusting. Nobody needs to do that. So that's a block, I think, it was certainly a block for me...,I hated it. Whereas I liked smoking. Whereas I like using that in a way, I mean, you know, it's a satisfying thing to do, a pleasant thing to do (11K, e-cigarette user)

E-cigarettes users in my study used e-cigarettes for variable reasons, they mentioned the stress relief that e-cigarettes gave them, keeping the rituals of smoking, the ability to smoke indoors, the absence of smell and bad taste. E-cigarettes were perceived as a safer option than cigarettes, they appealed more to them than other Nicotine Replacement Therapies and they seemed to be cheaper and socially acceptable activity. However, the several risks that the uptake of e-cigarettes use raises imply the need for finding a balance between benefits and risks. Hence, beneficent regulation of e-cigarettes 'will depend upon whether regulators see their goal as fostering choice and reducing harm from smoking or eliminating all nicotine use' (Hall et al. 2015: 3). UK regulators seem to be shifting towards achieving the former goal (i.e. fostering choice and reducing harm from smoking). PHE differentiated between smokers and nicotine users by stating, 'if nicotine delivery can be improved to mimic that of cigarettes more closely, these products have the potential mass appeal to challenge the primacy of smoked tobacco as the product of choice for nicotine users' (PHE 2014: 9). PHE also recommended 'encouraging smokers who cannot or do not want to stop smoking to switch to EC [e-cigarettes] as this could help reduce smoking related disease, death and health inequalities'(PHE 2015: 4) This conforms to the principle of justice and 'the purpose of maximising public utility' (Beauchamp and Childress 2009: 244). Allowing e-cigarettes use preserve justice as all the possible risks arising from the provision of e-cigarettes will be continually reported, published and addressed, while equal access to a product that may be helpful for any smoker is not restricted.

UK regulators have departed from imposing a ban on e-cigarettes. From May 2016, an e-cigarette which contains less than 20 mg/ml of nicotine and that has not sought medicinal regulations comes under the European Commission's revised Tobacco Products Directive (TPD) and will be regulated as a consumer product. However, an e-cigarette containing more than 20 mg/ml of nicotine or which makes smoking cessation claims is prohibited unless it has a medicinal license by the MHRA (ASH 2016). Although the new regulations may seem to fit within the principlism framework. They have been criticised for not respecting the autonomy of individuals who would like to use e-cigarettes containing more than 20 mg nicotine recreationally (Beard 2015). This new regulation of nicotine concentration, according to the European Commission, was based on their views of 'nicotine's classification as a toxic substance' (European Commission 2014: 5). This stance contradicts the accepted views, in the UK, where nicotine is seen as 'not carcinogenic', and where pure nicotine products, like NRT, were approved to be used without restrictions. Also, it is worth mentioning that although, e-cigarettes are not included in the smoke-free law, several organisations in the UK have banned e-cigarettes use in their premises (Bauld and Munafó 2015) which constitutes another deviation from the principlism framework. However, the new regulations, I argue, have the potential to instigate a social structural transformation of our perceptions of nicotine addiction as I will discuss next.

6   E-Cigarettes and the Social Transformation of Defining Nicotine Addiction

Although similar to Nicotine Replacement Therapy, e-cigarettes can be perceived as 'good' '(health-promoting)' nicotine, that cure tobacco addiction, their use as smoking substitute, shifts them towards 'bad' '(health destroying)' nicotine (Keane 2013: 190). Therefore, it is argued, e-cigarettes have exposed the 'artificial boundaries' between 'good' and 'bad' nicotine (Bell and Keane 2012: 246). E-cigarettes have merged both the 'good' and 'bad' nicotine and hence they are potentially facilitating a transformation of the circulated 'revulsive' views on nicotine addiction.

Regulations are considered an important factor in the creation of social acceptability. In their study, Sherrat et al. (2016) suggested that regulating e-cigarettes may be viewed as a social endorsement of e-cigarettes for some participants who felt uncertain or concerned about e-cigarettes safety. The recent advances in the UK regulations towards e-cigarettes imply a shift towards an environment of accepting the recreational use of nicotine in the form of e-cigarettes. Lessig (1995: 1030) argued that the anti-smoking regulations had stemmed from the 'social meaning of science', which proved that smoking is harmful. These regulations, Lessig believed, had stimulated the 'cultural redefinition of smoking', so smoking became socially unacceptable. I argue that half a century on, we could be seeing a sign of a cultural redefinition of addiction, again, stemming from the social meaning of science, which suggests that nicotine is not harmful. This redefinition may create a new social milieu and legal climate in which e-cigarettes become more acceptable and more accessible. Hence forming a new 'social norm' (Zhang et al. 2010). The e-cigarette is described as a device that 'meets many of the criteria for an ideal tobacco harm-reduction product' (RCP 2016: 63). Hence, e-cigarette may produce a new realm where consuming nicotine, in a form that resembles smoking, can coexist with public health agenda and their historical stance, which has been comprehensively against the visibility of any smoking like behaviour, and is sceptical about coinciding pleasure with health (Bell and Keane 2012: 245, 246; de Andrade et al. 2013). Consequently, not only normalising nicotine use but also normalising the act of using nicotine in a smoking like behaviour. E-cigarettes, I argue resulted in problematizing tobacco smoking rather than nicotine consumption; the cause of concern is no longer nicotine addiction, but is the addiction to tobacco smoking.

7 Conclusion

This chapter explored different ethical aspects raised as a result of e-cigarettes emergence. It reflected upon the considerable debate over the ethical issues of nicotine addiction, anti- smoking policies and the Harm Reduction approach. The chapter discussed some ethical dilemmas that e-cigarettes have created.

The data from the interviews highlighted issues of the controversy that surrounds e-cigarettes use among a sample of Stop Smoking Advisors and e-cigarettes users in SE England. This data provides 'local worlds of experience' of e-cigarettes and feeds into the wider bioethical debate. E-cigarettes have established a new entity encompassing 'clean' and 'dirty' nicotine. E-cigarettes may result in a social structural transformation of addiction; where a culture of nicotine use in this new form becomes socially acceptable in the West. Surely, this will instigate new moral challenges. Within the principlism framework, e-cigarettes can be permitted to be sold in ways that respect the autonomy of smokers by allowing them to reduce the harms of smoking. At the same time, in line with the non-maleficience principle and to maximise utility in the society, regulations such as restricting e-cigarettes advertisement and age restriction are put in place to protect younger generations from developing a new addiction. However, although UK regulators approved the safety of nicotine use in the form of Nicotine Replacement Therapy and licensed e-cigarettes, the justification for the cutoff point for the concentration of nicotine in e-cigarettes is questionable and still subject to further moral debate. E-cigarettes have challenged public health agendas by delivering a safer, than tobacco smoking, form of nicotine accompanied by smoking-like, pleasurable habit. Not only it exposed the artificial boundaries between 'good' and 'bad' nicotine, it made it fuzzier and hence will bring with it more moral challenges for bioethicists.

References

Aceijas, C. 2012. The Ethics in Substitution Treatment and Harm Reduction. An Analytical Review. *Public Health Reviews* 34 (1): 16.

Action on Smoking and Health (ASH). 2004. *Review of the Implementation of the Tobacco Product Regulation Directive 2001/37/EC*. http://www.ash.org.uk/files/documents/ASH_164.pdf.

———. January 2013. *Electronic Cigarettes*. Available at: http://www.sfata.org/wp-content/uploads/2013/06/ash.org_.uk_files_documents_ASH_715.pdf. Accessed 20 May 2014.

———. October 2014. *Q&A: The Regulation of Nicotine Containing Products in the UK*. Available at: http://ash.org.uk/files/documents/ASH_938.pdf. Accessed 1 Nov 2015.

———. 2016. *Use of Electronic Cigarettes (Vapourisers) Among Adults in Great Britain*. Fact Sheet. Available at: http://www.ash.org.uk/files/documents/ASH_891.pdf. Accessed 25 May 2016.

Adkison, S.E., R.J. O'Connor, M. Bansal-Travers, A. Hyland, R. Borland, H.H. Yong, K.M. Cummings, A. McNeill, J.F. Thrasher, D. Hammond, and G.T. Fong. 2013. Electronic Nicotine Delivery Systems: International Tobacco Control Four-Country Survey. *American Journal of Preventive Medicine* 44 (3): 207–215.

Bauld, L. 2016. *Expert Reaction to Study of E-Cigarette Use and Cigarette Smoking in Children*. Available at: Expert Reaction to Study of E-Cigarette Use and Cigarette Smoking in Children |Science Media Centre|Expert Reaction to Study of E-Cigarette Use and Cigarette Smoking in Children |Science Media Centre. Accessed 24 Feb 2016.

Bauld, L., and M.L. Munafó. 2015. *E-Cigarette Use in Enclosed Public Places: How Can Research Inform Regulation?* Available at: http://www.e-cigarette-summit.com/files/2015/11/930-Linda-Bauld1.pdf. Accessed 20 Nov 2015.

Bauld, L., K. Angus, and M. de Andrade. 2014. *E-Cigarette Uptake and Marketing!* Report No. 2014079. Available at: https://www.gov.uk/government/uploads/system/uploads/attachment_data/file/311491/Ecigarette_uptake_and_marketing.pdf. Accessed 30 Apr 2015.

Bean, S. 2014. *E-Cigarettes: Exploring Associated Ethical and Policy Implications for Hospitals* UPDATE FALL 30 (3). The Lung Association Ontario Respiratory Care Society. Available at: http://www.on.lung.ca/document.doc?id=2403. Accessed 23 Mar 2015.

Beard, A. 2015. Relapse Prevention Thursday, 5 November 2015. Available at: http://alanbeard.blogspot.com.es/2015/11/relapse-prevention.html?m=1. Accessed 20 Nov 2015.

Beauchamp, T.L., and J.F. Childress. 2001. *Principles of Biomedical Ethics*. 5th ed. New York: Oxford University Press.

———. 2009. *Principles of Biomedical Ethics*. 6th ed. New York: Oxford University Press.

Bell, K., and H. Keane. 2012. Nicotine Control: E-Cigarettes, Smoking and Addiction. *International Journal of Drug Policy* 23: 242–247. https://doi.org/10.1016/j.drugpo.2012.01.006.

Berridge, V. 1999. Histories of Harm Reduction: Illicit Drugs, Tobacco, and Nicotine. *Substance Use & Misuse* 34 (1): 35–47.

Biener, L., and J. Hargraves. 2014. A Longitudinal Study of Electronic Cigarette Use Among a Population-Based Sample of Adult Smokers: Association with Smoking Cessation and Motivation to Quit. *Nicotine Tobacco Research* 17 (2): 127–133.

Borry, P., P. Schotsmans, and K. Dierickx. 2005. The Birth of the Empirical Turn in Bioethics. *Bioethics* 19 (1): 49–71.

Buchanan, D.R. 2008. Autonomy, Paternalism, and Justice: Ethical Priorities in Public Health. *American Journal of Public Health* 98 (1): 15–21. https://doi.org/10.2105/AJPH.2007.110361.

Buchman, D.Z., W. Skinner, and J. Illes. 2010. Negotiating the Relationship Between Addiction, Ethics, and Brain Science. *AJOB Neuroscience* 1 (1): 36–45 http://www.ncbi.nlm.nih.gov/pmc/articles/PMC2910924/#R27.

Burstyn, I. 2014. Peering Through the Mist: Systematic Review of What the Chemistry of Contaminants in Electronic Cigarettes Tells Us About Health Risks. *BMC Public Health* 14 (1): 1.

Cahn, Z., and M. Siegel. 2011. Electronic Cigarettes as a Harm Reduction Strategy for Tobacco Control: A Step Forward or a Repeat of Past Mistakes? *Journal of Public Health Policy* 32 (1): 16–31.

Callahan, D. 1999. The Social Sciences and the Task of Bioethics. *Daedalus* 128: 275–294.

Campbell, N. 2012. Medicalization and Biomedicalization: Does the Diseasing of Addiction Fit the Frame? In *Critical Perspectives on Addiction*, Advances in Medical Sociology, ed. J. Netherland, vol. 14, 3–25. Bingley: Emerald.

Caponnetto, P., D. Campagna, G. Papale, C. Russo, and R. Polosa. 2012. The Emerging Phenomenon of Electronic Cigarettes. *Expert Review of Respiratory Medicine* 6 (1): 63–74.

Czogala, J., M.L. Goniewicz, B. Fidelus, W. Zielinska-Danch, M.J. Travers, and A. Sobczak. 2014. Secondhand Exposure to Vapors from Electronic Cigarettes. *Nicotine & Tobacco Research* 16 (6): 655–662.

Dackis, C., and C. O'Brien. 2005. Neurobiology of Addiction: Treatment and Public Policy Ramifications. *Nature Neuroscience* 8 (11): 1431–1436.

Dar, R., and H. Frenk. 2004. Do Smokers Self-Administer Pure Nicotine? A Review of the Evidence. *Psychopharmacology* 17 (3): 18–26. https://doi.org/10.1007/s00213-004-1781-2.

de Andrade, M., & G. Hastings 2013. *Research Priorities and Policy Directions. Tobacco Harm Reduction and Nicotine Containing Products*. Cancer Research UK. Available at: http://www.cancerresearchuk.org/prod_consump/groups/cr_common/@nre/@pol/documents/generalcontent/tobacco-harm-reduction.pdf. Accessed (17 Dec 2013.

de Andrade, M., G. Hastings, and K. Angus. 2013. Promotion of Electronic Cigarettes: Tobacco Marketing Reinvented. *BMJ* 347: f7473. https://doi.org/10.1136/bmj.f7473.

Department of Health (DOH). 1998. *Smoking Kills: A White Paper on Tobacco*. Available at: http://www.archive.official-documents.co.uk/document/cm41/4177/contents.htm. Accessed 30 Dec 2013.

———. 2011. *Healthy Lives, Healthy People: A Tobacco Control Plan for England*. London: Her Majesty's Stationery Office. Available at: https://www.gov.uk/government/uploads/system/uploads/attachment_data/file/213757/dh_124960.pdf. Accessed 5 Jan 2014.

Doll, R. 2004. Evolution of Knowledge of the Smoking Epidemic. In *Tobacco and Public Health: Science and Policy*, ed. P. Boyle, N. Gray, J. Henningfield, J. Seffrin, and W. Zatonski, 3–16. New York: Oxford University Press.

European Commission. 2014. Memo Questions & Answers: New Rules for Tobacco Products. Available at: http://europa.eu/rapid/press-release_MEMO-14-134_en.htm. Accessed 20 Nov 2015.

Evans, J.H. 2000. A Sociological Account of the Growth of Principlism. *Hastings Center Report* 30 (5): 31–39.

Fairchild, A.L., R. Bayer, and J. Colgrove. 2014. The Renormalization of Smoking? E-Cigarettes and the Tobacco "Endgame". *The New England Journal of Medicine* 370 (4): 293–295.

Farsalinos, K., G. Romagna, D. Tsiapras, S. Stamatis Kyrzopoulos, and V. Voudris. 2013. Evaluating Nicotine Levels Selection and Patterns of Electronic Cigarette Use in a Group of "Vapers" Who Had Achieved Complete Substitution of Smoking. *Substance Abuse: Research and Treatment* 7: 139–146.

Farsalinos, K.E., G. Romagna, D. Tsiapras, S. Kyrzopoulos, and V. Voudris. 2014. Characteristics, Perceived Side Effects and Benefits of Electronic Cigarette use: A Worldwide Survey of More Than 19,000 Consumers. *International Journal of Environmental Research and Public Health* 11 (4): 4356–4373.

Fox, R. 1976. Advanced Medical Technology – Social and Ethical Implications. *Annual Review of Sociology* 2: 231–268.

Gartner, C.E., W.D. Hall, S. Chapman, and B. Freeman. 2007. Should the Health Community Promote Smokeless Tobacco (Snus) as a Harm Reduction Measure? *PLoS Medicine* 4 (7): e185.

Glover, E.D. 2006. Successfully Treating Nicotine Dependence. *American Journal of Health Education* 37 (1): 6–14. https://doi.org/10.1080/19325037.2006.10598872.

Goniewicz, M., J. Knysak, M. Gawron, L. Kosmider, A. Sobczak, J. Kurek, A. Prokopowicz, et al. 2014. Levels of Selected Carcinogens and Toxicants in Vapour from Electronic Cigarettes. *Tobacco Control* 23: 133–139. https://doi.org/10.1136/tobaccocontrol-2012-050859.

Haimes, E. 2002. What Can the Social Science Contribute to the Study of Ethics? Theoretical, Empirical and Substantive Considerations. *Bioethics* 16 (2): 89–113.

Hajek, P. 2015. The Development and Testing of New Nicotine Replacement Treatments: From 'Nicotine Replacement' to 'Smoking Replacement'. *Addiction* 110: 19–22.

Hajek, P., J.F. Etter, N. Benowitz, T. Eissenberg, and H. McRobbie. 2014. Electronic Cigarettes: Review of Use, Content, Safety, Effects on Smokers and Potential for Harm and Benefit. *Addiction* 109 (11): 1801–1810.

Hall, W., C. Gartner, & C. Forlini. 2015. Ethical Issues Raised by a Ban on the Sale of Electronic Nicotine Device. *Addiction*. Article first published online: 5 Apr 2015. https://doi.org/10.1111/add.12898.

Hedgecoe, A. 2004. Critical Bioethics: Beyond the Social Science Critique of Applied Ethics. *Bioethics* 18 (2): 120–143.

Henningfield, J., and N. Benowitz. 2004. *'Pharmacology of Nicotine Addiction', Tobacco and Public Health: Science and Policy*, 129–147. Oxford: Oxford University Press.

———. 2010. Bioethics and the Reinforcement of Socio-technical Expectations. *Social Studies of Science* 40 (2): 163–186. https://doi.org/10.1177/0306312709349781.

Hoffmaster, B. 1992. Can Ethnography Save the Life of Medical Ethics? *Social Science & Medicine* 35: 1421–1431.

Hummel, K., C. Hoving, G.E. Nagelhout, H. de Vries, B. van den Putte, M.J. Candel, R. Borland, and M.C. Willemsen. 2015. Prevalence and Reasons for Use of Electronic Cigarettes Among Smokers: Findings from the International Tobacco Control (ITC) Netherlands Survey. *International Journal of Drug Policy* 26 (6): 601–608.

Jarvie, J.A., and R.E. Malone. 2008. Children's Secondhand Smoke Exposure in Private Homes and Cars: An Ethical Analysis. *American Journal of Public Health* 98 (12): 2140–2145.

Jensen, K.K. 2002. The Moral Foundation of the Precautionary Principle. *Journal of Agricultural and Environmental Ethics* 15: 39–55.

Keane, H. 2013. Making Smokers Different with Nicotine: NRT and Quitting. *International Journal of Drug Policy* 24 (3): 189–195.

Kleinman, A. 1999. Moral Experience and Ethical Reflection: Can Ethnography Reconcile Them? A Quandary for 'the New Bioethics'. *Daedalus* 128 (4): 69–97.

Lessig, L. 1995. The Regulation of Social Meaning. *University of Chicago, Law Review* 62 (3): 943–1045.

Levine, H.G. 1985. The Discovery of Addiction: Changing Conceptions of Habitual Drunkenness in America. *Journal of Substance Abuse Treatment* 2 (1): 43–57.

Lieber, S., and J. Millum. 2013. Preventing Sin: The Ethics of Vaccines against Smoking. *Hastings Center Report* 43: 23–33. https://doi.org/10.1002/hast.159.

Marcotte, A. 2014. The Moral Panic Over E-Cigarettes Intensifies. *Slate Blog*, 5 March. Available at: http://www.slate.com/blogs/xx_factor/2014/03/05/e_cigarettes_are_all_the_rage_with_kids_these_days_what_else_to_do_but_panic.html. Accessed 10 Apr 2015.

Markou, A. 2008. Neurobiology of Nicotine Dependence. *Philosophical Transactions of the Royal Society of London. Series B Biological sciences* 363: 3159–3168. https://doi.org/10.1098/rstb.2008.0095.

McCarthy, J. 2003. Principlism or Narrative Ethics: Must We Choose Between Them? *Medical Humanities* 29 (2): 65–71.

McNeill, A., L.S. Brose, R. Calder, S.C. Hitchman, P. Hajek, and H. McRobbie. 2015. *E-Cigarettes: An Evidence Update*. Public Health England. Available at: https://www.gov.uk/government/publications/e-cigarettes-an-evidence-update. Accessed 2 Mar 2016.

Medicines and Healthcare Products Regulatory Agency (MHRA). 2010. *The Use of Nicotine Replacement Therapy to Reduce Harm in Smokers*. Available at: https://www.gov.uk/drug-safety-update/nicotine-replacement-therapy-and-harm-reduction. Accessed 20 Apr 2016.

Mill, J.S. 1859. On Liberty. Available at: https://ebooks.adelaide.edu.au/m/mill/john_stuart/m645o/contents.html. Accessed 20 Apr 2015.

National Centre for Smoking Cessation and Training (NCSCT). 2016. *Electronic Cigarettes: A Briefing for Stop Smoking Services.* Available at: http://www.ncsct.co.uk/usr/pub/Electronic_cigarettes._A_briefing_for_stop_smoking_services.pdf. Accessed 15 Apr 2016.

National Institute for Health and Clinical Excellence (NICE). 2012. *Public Health Guidance on Tobacco Harm Reduction, Safety, Risk and Pharmacokinetics Profiles of Tobacco Harm Reduction Technologies Report.* Available at: http://www.nice.org.uk/nicemedia/live/14178/64034/64034.pdf. Accessed 1 July 2013.

————. 2013a. *Tobacco Harm-Reduction Approaches to Smoking: Guidance.* Available at: http://www.nice.org.uk/nicemedia/live/14178/63996/63996.pdf. Accessed 17 July 2013.

————. 2013b. *Nicotine Products Can Help People to Cut Down Before Quitting Smoking.* Available at: http://www.nice.org.uk/newsroom/news/NicotineProductsCanHelpPeopleToCutDownBeforeQuittingSmoking.jsp. Accessed 10 July 2013.

National Institute on Drug Abuse. 2008 *Drugs, Brains, and Behavior: The Science of Addiction,* revised edn. Washington, DC: National Institute on Drug Abuse.

National Institutes of Health U.S. 2007. *The Science of Addiction.* Department of Health and Human Services. Available at: https://www.drugabuse.gov/sites/default/files/sciofaddiction.pdf. Accessed 30 Apr 2015.

Newcombe, R. 1992. The Reduction of Drug Related Harm: A Conceptual Framework for Theory, Practice and Research. In *The Reduction of Drug Related Harm,* ed. P.A. O'Hare et al., 1–15. London: Routledge.

————. 2015. The Reduction of Drug-Related Harm: A Conceptual Framework for Theory, Practice and Research. International Use Substance Library (Drug text), Article 2, Addiction. Available at: http://www.drugtext.org/Various-general/the-reduction-of-drug-related-harm-a-conceptual-framework-for-theory-practice-and-research.html. Accessed 30 Apr 2015.

Orleans, T., and J. Slade. 1993. *Nicotine Addiction: Principles and Management.* Oxford: University Press.

Page, K. 2012. The Four Principles: Can They Be Measured and Do They Predict Ethical Decision Making? *BMC Medical Ethics* 13 (1): 1.

Parascandola, M. 2005. Lessons from the History of Tobacco Harm Reduction: The National Cancer Institute's Smoking and Health Program and the "Less Hazardous Cigarette". *Nicotine & Tobacco Research* 7 (5): 779–789.

Pepper, J., S. Emery, K. Ribisl, C. Rini, and N. Brewer. 2015. How Risky Is It to Use E-Cigarettes? Smokers' Beliefs About Their Health Risks from Using Novel and Traditional Tobacco Products. *Journal of Behaviral Medicine* 38 (2): 318–326. https://doi.org/10.1007/s10865-014-9605-2.

Pope, T.M. 2000. Balancing Public Health Against Individual Liberty: The Ethics of Smoking Regulations. *University of Pittsburgh Law Review* 61 (2): 419–498.

Public Health England (PHE). 2014. Electronic Cigarettes: Reports Commissioned by PHE, May 2014. Available at https://www.gov.uk/government/uploads/system/uploads/attachment_data/file/311887/Ecigarettes_report.pdf. Accessed 20 Sept 2015.

————. (2015) *E-Cigarettes: An Evidence Update: A Report Commissioned by Public Health England.* Available at https://www.gov.uk/government/uploads/system/uploads/attachment_data/file/457102/Ecigarettes_an_evidence_update_A_report_commissioned_by_Public_Health_England_FINAL.pdf. Accessed 2 Oct 2015.

Read, J., N. Haslam, L. Sayce, and E. Davies. 2006. Prejudice and Schizophrenia: A Review of the 'Mental Illness Is an Illness Like Any Other Approach. *Acta Psychiatrica Scandinavica* 114: 303–318.

Richardson, A., J. Pearson, H. Xiao, C. Stalgaitis, and D. Vallone. 2014. Prevalence, Harm Perceptions, and Reasons for Using Noncombustible Tobacco Products Among Current and Former Smokers. *American Journal of Public Health* 104 (8): 1437–1444.

Roe, G. 2005. Harm Reduction as Paradigm: Is Better Than Bad Good Enough? The Origins of Harm Reduction. *Critical Public Health* 15 (3): 243–250.

Royal College of Physicians (RCP). 2007. *Harm Reduction in Nicotine Addiction: Helping People Who Can't Quit.* A Report by the Tobacco Advisory Group of the Royal College of Physicians. London, United Kingdom. Available at: http://www.rcplondon.ac.uk/sites/default/files/documents/harm-reduction-nicotine-addiction.pdf. Accessed 5 Dec 2013.

———. 2016 *Nicotine Without Smoke: Tobacco Harm Reduction.* A Report by the Tobacco Advisory Group of the Royal College of Physicians. Available at: https://www.rcplondon. ac.uk/projects/outputs/nicotine-without-smoke-tobacco-harm-reduction-0. Accessed 25 May 2016.

Russell, M.A. 1976. Low-Tar Medium-Nicotine Cigarettes: A New Approach to Safer Smoking. *British Medical Journal* 1 (6023): 1430–1433.

Russell, M. 1991. The Future of Nicotine Replacement. *British Journal of Addiction* 86 (5): 653–658.

Saletan, W. 2009. Vapor War Our Irrational Hostility Toward Electronic Cigarettes. *Slate Blog*, 3 June. Available at: http://www.slate.com/articles/health_and_science/human_nature/2009/06/vapor_war.2.html. Accessed 20 Apr 2015.

Samuel, G. 2014. *fMRI for Severely Brain Injured Patients: A Media Analysis.* Doctoral Dissertation, School of Social Sciences Theses.

Samuel, G., and C. Brosnan. 2011. Deep Brain Stimulation in Parkinsonian Patients: A Critique of Adopting the Principlism Framework of Bioethics as a Form of Ethical Analysis for the Decision-Making Process. *American Journal of Bioethics Neuroscience* 2 (1): 20–22.

Scragg, R., R.J. Wellman, M. Laugesen, and J.R. DiFranza. 2008. Diminished Autonomy over Tobacco Can Appear with the First Cigarettes. *Addictive Behaviors* 33 (5): 689–698.

Sherratt, F.C., L. Newson, M.W. Marcus, J.K. Field, and J. Robinson. 2016. Perceptions Towards Electronic Cigarettes for Smoking Cessation Among Stop Smoking Service Users. *British Journal of Health Psychology* 21 (2): 421–433.

Shickle, D. 2009. The Ethics of Public Health Practice: Balancing Private and Public Interest Within Tobacco Policy. *British Medical Bulletin* 91 (1): 7–22. https://doi.org/10.1093/bmb/ldp022.

The Lancet editorial. 2013. E-Cigarettes: A Moral Quandary. *The Lancet* 382 (9896): 914.

The President's Council on Bioethics. 2003. *Session 2: Medicalization: Its Nature, Causes, and Consequences, Discussion of a Correspondence Between Paul McHugh, M.D. and Leon R. Kass, M.D. THURSDAY, June 12, 2003.* https://bioethicsarchive.georgetown.edu/pcbe/. https://bioethicsarchive.georgetown.edu/pcbe/transcripts/jun03/session2.html. Accessed 23 Mar 2015.

Thomas, B.P., and L.O. Gostin. 2013. Tobacco Endgame Strategies: Challenges in Ethics and Law. *Tobacco Control* 22: i55–i57. https://doi.org/10.1136/tobaccocontrol-2012-050839.

Turner, L. 2009. Anthropological and Sociological Critiques of Bioethics. *Journal of Bioethical Inquiry* 6 (1): 83–98.

U.S. Department of Health and Human Services (DHHS). 1988. *The Health Consequences of Smoking: Nicotine Addiction.* A Report of the Surgeon General. Atlanta: U.S. Department of Health and Human Services, Public Health Service, Centers for Disease Control, National Center for Chronic Disease Prevention and Health Promotion, Office on Smoking and Health. 88-8406.

Verweij, M. 2009. Tobacco Discouragement: A Non-Paternalistic Approach. In *Ethics, Prevention, and Public Health*, ed. A. Dawson and V. Verweij, 179–197. Oxford: Clarendon Press.

Warner, K.E. 2013. An Endgame for Tobacco? Editorial. *Tobacco Control* 22: i3–i5. https://doi.org/10.1136/tobaccocontrol-2013-050989.

Wayne, H., L. Carter, and K.I. Morley. 2003. Addiction, Neuroscience and Ethics. *Addiction* 98 (7): 867–870. https://doi.org/10.1046/j.1360-0443.2003.00400.x.

West, R., and J. Brown. 2013. *Theory of Addiction.* Hoboken: Wiley.

West, R., J. Brown, and E. Beard. 2014. *Trends in Electronic Cigarette Use in England.* Smoking Toolkit. Available at: http://www.smokinginengland.info/latest-statistics/. Accessed 2 Dec 2015.

Williams, C., and S. Wainwright. 2010. Sociological Reflections on Ethics, Embryonic Stem Cells and Translational Research. In *Contested Cells: Global Perspectives on the Stem Cell Debate*, ed. B.J. Capps and A.V. Campbell. London: Imperial College Press.

Wise, R.A. 2000. Addiction Becomes a Brain Disease. *Neuron* 26: 27–33.

World Health Organisation (WHO). 2010. *Gender, Women, and the Tobacco Epidemic. 7. Addiction to Nicotine.* Available at: http://www.who.int/tobacco/publications/gender/en_tfi_gender_women_addiction_nicotine.pdf. Accessed 23 Mar 2015.

———. 2014. *Facts & Figures. Management of Substance Abuse. Tobacco.* Available at: http://www.who.int/substance_abuse/facts/tobacco/en/. Accessed 29 Sept 2014.

World Health Organisation Framework convention on Tobacco control (FCTC) (2014) *Electronic Cigarettes (E-Cigarettes) or Electronic Nicotine Delivery Systems.* FCTC/COP/6/10. Available at: http://apps.who.int/gb/fctc/PDF/cop6/FCTC_COP6_14-en.pdf. Accessed 22 Oct 2015.

Zhang, X., D.W. Cowling, and H. Tang. 2010. The Impact of Social Norm Change Strategies on Smokers' Quitting Behaviours. *Tobacco Control* 19 (Suppl 1): i51–i55.

Performing Risk & Ethics in Clinicians' Accounts of Stem Cell Liver Therapies

Steven Wainwright, Mike Michael, and Clare Williams

1 Introduction

Stem cells have huge potential in the fields of tissue engineering and regenerative medicine as, in principle, they hold the capacity to produce every type of cell and tissue in the body. They therefore arguably promise a medical revolution in the treatment of diverse areas, including cardiovascular and neurodegenerative diseases, and the possibility of the replacement of organ transplants with stem cell transplants (Scott 2006). In this paper we consider the emergent possibilities for the use of stem cell therapies to treat liver disease. Needless to say, realizing such possibilities is a fraught business, not least because there are a range of risks that can be associated with the procedures and processes of stem cell therapies. However, such risks pertain not simply to the risks faced by patients, but to a complex nexus of risks that range from the 'simple' risks of wasting valuable liver cells through to the 'meta-risks' of failing to take risks. This suggests an approach to risk that is 'performative', in which 'risk' is not so much a matter of more or less technical assessments of costs and benefits, or a reflection of a particular cultural perspective, but rather a resource in dynamically accounting for multiple institutional, medical and social dilemmas.

In this paper we set out to explore these enactments of risk by clinicians involved in the development of stem cell therapy for liver disease. In the process, we contribute to a performative re-thinking of how 'risk' can be analytically treated in relation to health. As such, we begin with an outline of the key scientific, medical and social science literature on stem cells. We then go on to summarise our theoretical thinking

S. Wainwright · C. Williams (✉)
Department of Social and Political Sciences, Brunel University London, London, UK
e-mail: Steven.wainwright@brunel.ac.uk; Clare.williams@brunel.ac.uk

M. Michael
Department of Sociology, Philosophy and Anthropology, University of Exeter, Exeter, UK
e-mail: m.michael@exeter.ac.uk

© Springer International Publishing AG, part of Springer Nature 2018
H. Riesch et al. (eds.), *Philosophies and Sociologies of Bioethics*,
https://doi.org/10.1007/978-3-319-92738-1_9

around the analysis of risk. After a brief note on methodology, we come to the bulk of the paper where, drawing on interview data, we explore the complex enactment of risk by clinicians in the context of emergent stem cells therapies (Williams et al. 2003; Brown and Michael 2002). In conclusion, we briefly consider some of the implications of our analysis for the further analysis of the 'enactments of risk' in medical and biomedical settings.

2   Stem Cell Revolutions?

Stem cells may herald a medical revolution in the treatment of cardiovascular disease; neurodegenerative diseases, such as Alzheimer's and Parkinson's; diabetes; and even cancer (Alison et al. 2006). This vision of a stem cell based regenerative medicine is a recent phenomenon, and stems from 1998 when two seminal papers described the growth *in vitro* of human embryonic stem (hES) cells derived either from the inner cell mass (ICM) of the early blastocyst (Thomson et al. 1998) or the primitive gonadal regions of early aborted fetuses (Gearhart 1998). Stem cell biology has evolved briskly over the last few years and it is one of the most rapidly developing areas within the life sciences. The two defining features of ES cells are *self-renewal*, where a few cells can potentially produce large numbers of daughter cells, and *differentiation*, where unspecialised cells are transformed into an array of different mature cell types (Lanza et al. 2004). In principle, stem cells can produce every type of cell and tissue in the body and because they can transform into any of the around 220 cell types that make up the human body they are arguably the most promising approach in the fields of cell transplantation and regenerative medicine (Gearhart 2005). However, the difficulties of laboratory work on stem cells is vividly illustrated by a gap of 22 years between the isolation of stem cells in the mouse (Evans and Kaufman 1981), and the first culture of a human embryonic stem (hES) cell line in the UK (Pickering et al. 2003). The prospects for liver cell transplantation offer a key way to examine the trajectory of stem cell therapy from the laboratory to the clinic and, potentially, to the commercial market place (DOH 2005).

Social researchers have explored numerous barriers and opportunities critical to the development of embryonic, fetal and adult stem cell research/treatment. This is partly because it is helpful to have grounded information about the factors predisposing to successful diffusion and uptake of specific stem cell 'products', and conversely, the barriers to this process. By mapping the range of meanings attributed to stem cells, social research extends analytical understandings of how they are constituted in specific settings. This paper explores the shifting significance of stem cells in the field of liver disease, thereby revealing some of the ways in which narratives about stem cells are produced, resisted, negotiated and accommodated. Since 1998, when human embryonic stem cell lines were first isolated a series of social research papers have begun to map the key debates in the burgeoning area of stem cell research (Franklin 2001; Waldby 2002; Parry 2003; Williams et al. 2003; Franklin 2005; Kitzinger and Williams 2005). However, these examples of social science all utilise documentary sources. In contrast, our research, elements of which we draw upon in this paper, employs interviews with stem cell scientists. We have discussed the prospects and problems of human Embryonic Stem (hES) cell research and cell transplantation in a series of publications on expectations, the body, and ethics which draw upon the perceptions of scientists (Wainwright et al. 2006a, b, c, 2007; Michael et al. 2007a, in 2007b; Williams et al. 2008). In this paper, however, we turn to a discussion of how clinicians view risk in interrelated fields of stem cells and liver cell therapy.

3 From Judging to Enacting Risk

It goes without saying that 'risk' is a complex and evolving construct (e.g. Lupton 1999) and in the present case, we can only focus on particular facets. In late modernity, risk is a key motif in the assessment of biomedical innovation (see Rose 2007) and arguably part and parcel of mundane accounts of everyday life (Beck 1992; Tulloch and Lupton 2003). In the present instance, we might be tempted to suggest that our clinicians' accounts of the risks associated with stem cell therapies for liver disease straddle these (not altogether clearly demarcated – see Hamilton et al. 2007) domains of lay and expert discourses around risk. Clinicians must concurrently draw on 'expert' techniques (for example, embodied in ethical consideration) and make 'folk' judgements (for instance, about the likelihood of therapeutic 'success' in specific clinical cases). However, this does not quite capture the complexity of these risk assessments. To be sure, as we shall see, clinicians must be able to warrant their decisions in terms of technical guidelines of 'risk assessment' or 'cost-benefit analysis', but this is not always seen in a positive light. Indeed, clinicians also value a more liberal regime toward risk-taking. This latter 'courage to fail' (see below) can be read in terms of Douglas's radically Durkheimian grid/group theory in which people's perceptions of risk are seen to serve in the reproduction of the internal structure of the group to which individuals belong, and its relation to other collectivities (e.g. Douglas 1970; Schwarz and Thompson 1990; but see Johnson 1987). Without going into details, certain 'collectivities' (e.g. low grid, low group) can be regarded as individualist and entrepreneurial with a commitment to a risk-oriented ethos of the 'bottom line' and 'bold experimentation'. Superficially, this might be applied to our clinicians, and indeed Rayner (1986) found just this sort of 'self-confidence' amongst consultants in the management of radiation hazards in a hospital.

However, this perspective tends to downplay the performative dimension of 'doing', and accounting for, risk-taking. In keeping with now classic work on the role of accounts or enactments in the 'making' of particular versions the social and material world (see, for instance, Mol 2002), we also want to explore risk in terms of how clinicians are enacting not only a version of their own 'group', but also versions of other groups such as patients, regulators, experimental scientists and the public more generally (as well as 'liver disease' and 'stem cells', of course). In other words, we aim to study how clinicians tacitly represent and address a multiplicity of constituencies, and thereby serve in their (partial) 'making' (see Hacking 1986) where certain possibilities of action are enabled, while others are disabled. On this score, we might better consider 'risk' as a resource through which to address and 'make' such constituencies. By way of brief example, clinicians' accounts of risk-taking clinical research reflect not simply a sort of 'bravado' in relation to the driving knowledge and skills forward (e.g. in the use of stem cell transplantation to treat acute liver disease), but are also a response to the perceived expectations of other constituencies which are thereby enacted in a particular way. That is to say, the taking and accounting of risks generates what Brown and Michael (2002) have called 'meta-risks' – how one makes judgements about risk and uncertainty for this or that procedure leads to risks in relation to this or that audience (e.g. regulators, publics). As a result, how one can act in relation to those constituencies reflects the way they have been 'made', but also that 'making' affects how those constituencies can themselves warrantably act.

In summary, we take a performative perspective on the way that our clinicians address the risks associated with stem cell therapies as applied to liver disease. We attempt to draw attention to the ways their accounts enact particular versions of other constituencies, but we also locate the clinicians themselves within a complex nexus of institutional, medical and social factors.

4   Note on Methods

In this paper we use data from two ESRC funded projects on the prospects and problems of translational research in the field of stem cell research where we conducted over 60 in-depth interviews with scientists and clinicians in some of the leading labs and clinics in the UK and the USA, exploring their views on the bench-bedside interface in the fields of neuroscience, diabetes and liver disease. Here the focus is on interviews with five Consultant clinicians (three are Professors) who run a leading cell transplant programme. The interviews lasted between 1 and 2 h, took place in the experts' offices, and with permission were taped and transcribed. Open-ended questions and an informal interview schedule were used in order to encourage scientists and clinicians to speak in their own words about their experiences. Transcripts were analysed by content for emergent themes (Weber 1990), which were then coded (Strauss 1987). The research team discussed the data and analysis which enabled different perspectives to be incorporated, and added to the richness of the analysis.

5   Stem Cells and Liver Disease

The growing shortage of organs for transplant has helped fuel a burgeoning interest in the prospects for stem cell transplantation (Burns et al. 2004). From this viewpoint, cell transplants may potentially replace the entire field of organ transplantation with an array of stem cell therapies (Donovan and Gearhart 2001). In the field of liver disease 'cell therapy' may take several forms: for example, as a bridge to liver transplantation (Seldon and Hodgson 2002), as the basis for an artificial liver support system (Plevris and Hayes 2001), as a means to induce immunological tolerance in transplant patients (Norman and Turka 2001), and as cell transplant therapy which can already cure some genetic liver diseases in children (Mitry and Dhawan 2002). However, all of these strategies are experimental and those working in the field of 'liver cell therapies' contest the prospects for any of these approaches. We will explore some of these divergent views shortly, but before we do so we want to turn briefly to the disputed nature of 'what counts as a liver stem cell'.

The very nature of stem cells is scientifically contested (Blau et al. 2001). One of the key debates here is whether stem cells are best thought of as an entity or as a function. Moreover, even when stem cells are seen as an entity their status as an entity is still contested. For instance, one paper argues that there are at least eight types of 'liver stem cells', and that almost all the existing scientific literature elides these important distinctions (Dahlke et al. 2004). Moreover, and to make things even more complex, each of these eight different 'liver stem cell entities' has differing functions. Obviously, these differences in terminology and different ways of seeing a fairly specific aspect of the biological (and social) variety of (liver) stem cells is very important when reviewing research on stem cells in hepatology (Grompe 2004), although which of these liver stem cells 'is best' is not yet known (Fausto 2004). If this smacks of biological reductionism it should be noted that Theise (2003), a leading liver stem cell researcher, argued that stem cell research is reversing the 'biological reductionism' that has characterised the history of biomedical research and that biomedical scientists are entering a phase when tissue biology is becoming more important than molecular biology. In our view, given the amount of resources spent on molecular biology, this may be more of a rhetorical flourish than an empirical observation of the nature of modern biomedical science (Jasanoff 2005). However, this link between cells, tissues and organs will serve as a useful way to introduce our discussion of the problems and prospects of liver cell – from now onwards, hepatocyte – transplants and therapies.

Liver Cells, Cell Transplantation and Risk

As we indicated above, there are now 'proof of concept' studies that suggest that hepatocyte transplants have therapeutic benefit, for example, in children with a life threatening genetic liver defect (Mitry and Dhawan 2002). We will use this example to explore some of the tensions both between the bench and the bedside, and the tensions within the competing medical teams that are developing innovative new treatments in this specialised field. In brief, should valuable clinical grade hepatocytes be used for scientific research, or should they be tried in experimental treatments? And if the answer is experimental treatments, then which of the competing experimental strategies of: hepatocyte transplants in children, cell transplants for (adult) patients with acute liver failure, or use of the cells in 'artificial liver support systems' in critical care patients, should be the focus of clinical research efforts? In what follows we address some of the issues around the shaping of risk and hepato-cyte transplants through four main themes: perspectives on the clinical use of liver cells; the clinical and scientific; experimental treatments; and regulation and profes-sional autonomy. We discuss how clinicians at the forefront of treatment for life threatening liver disease utilise such risk discourses.

5.1 Perspectives on the Clinical Use of Liver Cells

One of the key potential areas for human embryonic, fetal and adult stem cell research and treatment is that of therapies for liver disease. Currently, liver trans-plantation is the treatment of choice (O'Grady et al. 2006) but two major problems, the acute global shortage of organs available for transplant (Lock 2001) and the morbidity and mortality associated with immunosuppression (Forsythe 2001) have resulted in a search for alternative treatments. These include the transplantation of stem cells (Suknikh and Shitl 2003) and hepatocytes (liver cells) (Mitry and Dhawan 2002) to aid liver regeneration; xenotransplantation (Michael and Brown 2005); and the bioengineering of artificial livers (Plevris and Hayes 2001). Stem cells are seen as the key to success in all four of these strategies of research and treatment (Polak et al. 2002). Liver disease therefore offers a unique lens through which to examine the trajectory of 'stem cell therapy' from the laboratory to the clinic. However, in contrast to stem cells, hepatocytes are found only in the liver and are difficult to obtain in sufficient numbers, so 'the prospects for culturing, expanding and matur-ing liver stem cells are enormously exciting' (Seldon and Hodgson 2002: 165).

The use of liver cells in clinical research can usefully be divided into several categories. We begin with debates around the shift from organ transplants to cell therapies. We then turn to the use of liver cells in critical care settings, and discuss cell transplants as a novel treatment for Acute Liver Failure (ALF), and the rather fraught attempts to develop artificial liver support systems (Plevris and Hayes 2001). Next, we present what seems a more promising use of liver cells in our exploration of hepatocyte transplants in children with life threatening genetic disor-ders of liver metabolism. We conclude our overview of the field with some thoughts on the prospects of embryonic stem cells as 'a cure for liver disease' (Alison et al. 2000). In what follows, we draw on interviews with leading UK consultant physi-cians and consultant surgeons in the field of liver disease.

From Organ Transplants to Cell Therapies?

Organ transplantation has been one of the defining features of shifts from a medi-calised to a biomedicalised view of health problems in Western societies in the late twentieth century (Clarke et al. 2003). However, the ever-increasing demand for organs has outpaced supply and this has led to significant yearly increases in waiting times for organ transplants, and to increases in deaths of patients awaiting transplant (Norman and Turka 2001). One potential way to tackle this crisis is by the development of cell transplantation, where the required cells (cardiac, liver etc.) are made from another potentially unlimited source of cells like embryonic stem cells (Polak et al. 2002). This potential shift from organ transplants to cell transplants is one example of the rapid evolution of the transplant field. Inevitably, the dynamic nature of organ transplantation attracts a certain type of clinician, as a transplant surgeon recalls:

> So liver transplantation was a fast changing field, and it was good. In personal terms I found it exciting. I found it suited my personality. I like change, I like the challenge, I like being slightly uncomfortable. Surgeon 32.

These features of enjoying the risks associated with developing the transplant field are captured in autobiographical accounts of the development of liver transplantation (Calne 1998; Starzl 2003). Starzl performed the first 'successful' human liver transplant in the USA (Starzl et al. 1963), and Calne was the first to transplant the liver in humans in the UK (Calne and Williams 1968). The next quotation names both Roy Calne and Tom Starzl as the pioneer 'risk takers' of liver transplantation (see Calne 2003):

> Surgeons are so used to seeing results, looking at things like cure… And if you look you'll find that surgeons often have been the clinical leads. Roy Calne, Tom Starzl, people like that. You'll find that behind every medical innovation, there will often be a surgeon as well. Who took it forward – it was the clinicians [rather than scientists]. People had to be bold about it. Surgeon 46.

Driving innovation and being bold are here seen as part of the risky disposition required to be successful in the transplant field and those in the field see this, in turn, as vital for the development of organ and cell transplantation. There is a mutual shaping of individuals and institutions (Wainwright et al. 2006d). Further, tacit in this account is an enactment of broader constituencies – such as publics – who 'collude' in this risk-taking. Yet, as we shall see later, particular publics can also outstrip the clinicians in risk-taking. For all the clinical boldness on display, there are also important differences between organ and cell transplants that might compromise such daring:

> I mean the difference between cell transplantation and solid organ transplantation is, if you transplant a whole liver, it's very clear when the liver fails as the patient will die. With cell transplantation, the point at which the graft fails is less easily defined. Re-transplantation is a lesser thing. As a rule it's an infusion of more cells, and so it changes your concept of transplantation. Surgeon 32.

Here we see how success and failure, and thus risk itself, are measured in different ways between cell and organ transplants. In particular, risks of rejection have to be conceptualised and practiced in different ways. In the next sections we examine the various ways in which hepatocytes can be used as the basis for experimental, and potentially life-saving, therapies.

Cell Transplants for Acute Liver Failure in Adults

Most liver transplants are for chronic liver disease (Wainwright 1994, 1995, 1997). However, ALF (Acute Liver Failure) can be a life threatening condition where some patients can 'be saved' through liver transplantation (Sargent and Wainwright 2006). Once the patient has been transplanted then they are, effectively, chronically ill for life as they must take potent immunosuppressive drugs which often have unpleasant side effects (Wainwright and Gould 1997). The liver is a unique organ that is able to regenerate, and so if patients can be kept alive long enough to regain their normal liver function (by, say, recovering from the infection or poisoning that caused their ALF) there will be no need for the major surgery of liver transplantation. Both cell transplants and artificial liver support systems are experimental therapies that aim to give the critically ill ALF patient time to recover *without* an organ transplant. In fact, hepatocyte transplantation can be seen as the next step in the historical shifts from whole liver, to partial liver (one lobe), to auxiliary liver (a small piece of liver) to hepatocyte, transplants (Forsythe 2001). Making such shifts in therapy, however, requires expertise and this expertise is much more readily available in some centres than others:

> In acute liver failure, many centres don't see enough patients to develop the technical exper-
> tise to manage the patients. I think worldwide there's been some scepticism that the results
> of auxiliary transplantation are as good as whole liver transplantation, although we've pub-
> lished some studies on this. We're just about to submit our overall experience now. There is
> interest worldwide in what we do. People are willing to copy us. But it is a very specialised
> field. In the States there are over 150 liver transplant centres, many of which will do 10 or
> 20 liver transplants a year, which is unacceptable. And the concentration of expertise into a
> small number of centres doesn't happen in the same way as it perhaps does here [in the
> UK]. Surgeon 32.

In addition to problems with the acquisition of expertise and the subsequent diminishing of risk, a second set of major problems associated with the development of any cell based therapy for ALF is the huge number and volume of cells needed for therapy. This requirement creates logistical and monetary risks which act as barriers to treatment that currently seem insurmountable. In essence, the lab cannot supply the clinic with material even for small-scale experimental use of these cells for ALF:

> If you were to use stem cells to treat ALF [Acute Liver Failure], you're going to need an
> awful lot of them. 400 grams is what all of the estimates would suggest as to what you need.
> It's also seriously expensive. So you're going to have to try and show efficacy of that… You
> are looking to support liver function for acute liver failure, when you've got total wipe-out.
> So, in order for stem cells to work in that setting, you've got to be able to ring someone up
> and say, 'I've got someone with acute liver failure,' and have the cells there and then within
> the next 12 hours. Then you've always got to have cells on the shelf ready to go, which
> invokes a phenomenal cost. Like seriously vast. And it's one of the main reasons all the
> companies thus far have gone bankrupt, because of the cost of generating that, even in the
> face of a controlled trial. So the cost to any healthcare system is going to be vast to do that
> at the present time. And it's always going to be expensive because of the stringency that will
> be required for using cells clinically. Physician 37.

Here, 'boldness' becomes diffused by the sheer organizational difficulties of accumulating expertise, on the one hand, and cells on the other. To go beyond these would be to enter a state of 'recklessness'. Put another way, claims to 'boldness' presuppose a backdrop of appropriate resourcing that render any risk-taking 'good' as opposed to 'bad', justified as opposed to reckless.

Artificial Liver Support Systems

The first attempts to develop artificial liver support systems date from around a decade after the successful introduction of haemodialysis machines for the treatment of renal failure (Cameron 2002). In the case of the kidney, however, a man-made filter is sufficient to mimic the key 'mechanical functions' of the human kidney. In contrast to the kidney, the liver is a much more complex metabolic organ and so any attempt at producing a 'liver machine' requires liver cells (Plevris and Hayes 2001). In the past, cells from pigs have been used in trials but more stringent 'xenotransplant guidelines' means that such an approach is no longer permitted in the UK (though animal cells are still used in some experimental liver machines elsewhere). Knowing which is the next 'winning technology' for liver disease is difficult for those in the field to be sure about, as Surgeon 32 explains:

> Because of our reputation people will often come to us if they've developed a new [liver support] machine that they want to validate. If people don't come to us, we'd be rather suspicious of them! And it's always difficult to know what is going to be the next sort of winning technology. What we're trying to do is develop interest in all the strands. So hopefully that will allow us to be in the right area. Surgeon 32.

In this quote, risk can be viewed in terms of curtailing options – failing to 'develop interest in all the strands'. This can also be understood in terms of prematurely opting for one particular development. The boldness entailed in such choice-making becomes problematic in a context where it is unclear what the next 'winning technology' is likely to be. Once again, boldness is diffused by the lack of resources – this time, that resource is 'knowledge'.

Moreover, there is a tension between different uses of 'clinical grade hepatocytes' (which currently are derived from human cadaver livers) that is evoked by a physician who argues that using them in liver support machines is a waste when such materials can be used for paediatric hepatocyte transplants:

> *With liver support machines a huge number of cells are needed to enable those to work. Do you see any way in which the sort of cells that you're using could be used for that?*

> If you use those cells in machines, that would be wasting your cells, because then they'll be dead after a few days. So if you have an unlimited supply of cells by genetic engineering or whatever you want to do, then it's different. But if somebody says that I have isolated cells from this donor and I want to put them in a reactor my view is that's not the best use of those cells. Physician 34.

For this clinician, risk is seen in terms of the most likely benefit to 'proximal' patients. A tacit cost-benefit assessment is rendered where the likely benefits of the 'clinical' use of the hepatocytes outweigh their experimental use in the machines. As such, immediate ostensible benefits trump long-term possible benefits to 'distal' patients. However, this account also tacitly enacts a public constituency that is more enthused by the likelihood of immediate benefits, and less concerned with the risky future possibilities of innovation (of support systems).

Hepatocyte Transplants for Children

Hepatocyte transplants for children are a very new medical technology, as the first successful transplants were only performed a few years ago (Mitry and Dhawan 2002). A leading surgeon explains how the idea of cell transplants grew out of ongoing treatment and research on auxiliary liver transplantation:

> We've done auxiliary liver transplant for the group with inborn errors of metabolism based in the liver who are non-cirrhotic. And because we found we could produce graft survival for more than ten years in these children, it developed our ideas of cell transplantation. And this is where the two meet together, because in Prader-Willi syndrome, you probably only need 1 or 2% functioning cell mass to produce a noticeable effect in the patient. So that's the point at which solid organ transplantation from an auxiliary transplant meets with the concept of cell transplantation. I think cell transplantation is part of the path. I don't think it's going to be the final treatment pathway. I'm sceptical that it will ever become THE viable way of treating people in the long term. But I think it's about learning, and we're learning a lot of different things along this way. Surgeon 32.

In the quotation above, we see how this surgeon regards this novel treatment as having a small role in the organ transplant field. The experimental nature of hepatocyte transplants also led a physician to argue for a low risk approach to any future expansion of this innovative therapy:

> I think at the moment we should probably be limited to doing it in people or children whose outcome without it is inevitably poor, learn more about it over a longitudinal timespan, and then it may become reasonable to go back and use it in healthier type people who are desirous of it. Physician 37.

Some of the ways of learning more about how cell transplants work over time, and the iterative nature of experimental clinical research and the subsequent 'diminishing of risk', are summarised in the following:

> Our ideas for the islets and hepatocytes transplants have come together in some respects, which are, how do you deliver sufficient cells, how do you put them in, in a viable state, how do you try and ensure there's good implantation, how do you ensure that there's adequate immunosuppression when you don't have a clear cut marker of graft function? You can demonstrate *improved* liver function, but you can't demonstrate *normalisation* of liver function. So this is probably because of the number of cells you can deliver into the liver. And I think that's one of the problems, it's getting above a certain volume. Surgeon 32.

Despite problems, such as infusing sufficient cells for a cure and ensuring that the cells infused survive 'for ever' rather than just for a period of months, clinicians working in the area espouse the usual transplant line that they must hold onto their 'courage to fail' attitudes and practices (see, Fox and Swazey 2002):

I think the key thing in hepatocyte transplants is that it's a very promising new territory. And the initial enthusiasm has probably died because of the longer-term outcome, but I don't think that should hold people back. Unless we keep pushing that, we're never going to find that fundamental link which will make the thing work forever. Until we make cell therapy work, and I'm sure it will work, we have to keep pushing the boundaries. Surgeon 46.

Pushing the boundaries here ostensibly comes from within a multidisciplinary team that is developing novel cell transplant therapies. Moreover, the experimental cell transplant programme has been 'piggy-backed' onto the large-scale liver (whole organ) transplant programme:

We are not under pressure from the surgeons who do liver transplants as we would only get the livers which they wouldn't use. So we are not competing for livers. If you were competing for livers, then it wouldn't go anywhere. The whole idea is that we use those livers for cell transplants that are not used by a standard transplant.

And how successful have your transplants been?

No one has been cured, but every patient has been benefited in his own way. Physician 34.

In this case, waste material in the form of 'spare pieces of liver' that are too small for even auxiliary transplant are being turned into clinical grade hepatocytes in a GMP (Good Medical Practice) cell isolation lab. Unfortunately, such clinical practice has not been used in some centres where hepatocyte transplants have been performed on humans:

But if something seriously goes wrong with the lab [for producing 'clinical grade' liver cells for transplant], then we would be in trouble. Then you can't safely transplant cells. The staff have to isolate these cells in a safe environment, it's crucial. Many places have done it without being in that environment! They've been lucky. Physician 34.

Here, the high risk approach of some other centres, who have transplanted 'research grade' hepatocytes into humans, is contrasted with the low risk 'clinical grade' transplants of the centre we studied. Even so, the overall research strategy is still risky, allied as it is to a pressure to 'push ahead' with new medical technologies:

If I'm not prepared to do something, someone else will do it. There's someone right behind me, who's going to clamber over my back, prepared to do that. There is great pressure on individuals and institutions to 'push ahead'. Surgeon 32.

In the above, we see the boundary between research and treatment, of what Renee Fox described in the 1960s and 1970s with the metaphors of the 'courage to fail' and the commitment to 'experiment perilous' (Fox and Swazey 2002) being recast in the more globalised setting of twenty first century biomedicine (Rajan 2006). In the context of a performative perspective on risk accounts, this risk-taking, in contrast to the 'safer options' preferred in the preceding section, while situated in relation to the research teams, also tacitly enact a wider public constituency which is more positive toward clinical research boldness (not least, perhaps, because of their perception of the beneficial outcomes of previous boldness in organ transplantation). In this case, it would seem that clinically high risk strategies are folded into socially low meta-risks.

Embryonic Stem Cells as 'A Cure for Liver Disease'?

In the four sections above we have outlined the competing claims for the development of different cell based therapies. What these discourses and practices have in common, however, is the need for an almost unlimited supply of cells. In this final section we therefore turn to the notion of ES cells as the best way to 'cure liver disease':

> My time frame is the next 10 or 15 years, and looking at interesting areas of research. And I think the cell transplant side clearly has limitations. They're apparent now. And you have to go to more pluripotent cells, and the far extreme of that is the embryonic stem cells which have huge potential, but maybe are less likely to behave well in appropriate environments. We don't really know. The question is, at what point do those two technologies meet? Surgeon 32.

Expectations of the cell transplant and ES cell technologies converging to produce a new era of regenerative medicine are seen here to be at least a decade away. This timeframe can, however, serve to enable both ES scientists and hepatocyte transplant clinicians to argue that 'more research' will enable the blending of these nascent medical technologies:

> My view is that there may be a day when somebody will give you a cupful of stem cells and then produce a cupful of liver stem cells from them. So my argument is that the current work we are doing on [transplanting] mature liver cells will help us to use that cup of stem cells when it comes. Because these mature cells need handling, they need monitoring. We are working at that now, so it's going to help build a platform for when the stem cell technology eventually arrives. Physician 34.

The notion of an emerging 'biomedical platform' is here based on the (expected) confluence of cell transplants and the transformation of ES cells into functioning cells. There is resonance here with the '40-year revolution' in cancer biology and clinical practice where FACS (Fluorescence Activated Cell Sorting) enabled the convergence of basic immunology with applied cancer medicine (see Keating and Cambrosio 2003). However, this example from the cancer field shows that it takes decades for a significant realignment in the nature of biomedical science to become commonplace, a theme which is explored from a social science perspective as 'the myth of the biotech revolution' (Nightingale and Martin 2004). From this point of view, the gradual evolution rather than revolution of cell transplants can be seen as mirroring previous innovations in biotechnology, and this is 'a good thing', as a surgeon involved in the field explains:

> What is clear with cell transplantation is that almost the worst possible thing that can happen is that it's an absolute success, because we'll have a technology we can't give to people because we don't have the raw materials. Surgeon 32.

Here, cell transplantation as a runaway success is, paradoxically, seen as a major risk that would undermine the development of the field. Furthermore, an unlimited supply of stem cells may be of little help to patients with the host of complications associated with chronic liver failure:

> The problem is that it's not just the lack of hepatocytes that causes patients to decompensate and get complications. It's usually sepsis, varices, ascites, and infection. We won't necessarily solve that by using stem cells! You might in a few of them, but not in all of them. When a disease has become chronic, it's more than just about supporting one cell line. It's supporting the consequences of that disease. Physician 37.

Here, various risks from the complex of symptoms of patients with liver disease may overwhelm any seemingly simple attempts at producing a cure through hepatocyte transplantation. In summary, the current consensus, in both the scientific and clinical literature, and from the experts in the field that we interviewed, is that stem cells (of whatever type) are likely to be of limited clinical use within the domain of liver disease:

> Embryonic stem cells I see as just a way of increasing our knowledge. I'm not convinced it's going to be the final answer. Surgeon 32.

Once more, we see how risk-taking is embedded in an assessment of the sorts of resources available. The clinical research risk-taking necessary for building the platform in the relatively dim and distant future might entail more social risks (in the sense of public disaffection and funding decreases) once the prohibitive levels of raw materials needed and the complications of liver disease are taken into account.

5.2   The Clinical and the Scientific

In this next section we will be exploring some of the cleavages that delimit the interrelated fields of science and medicine (Wainwright et al. 2006b). A Surgeon highlights some of the differences between the cultures of the lab and the clinic:

> To demonstrate the seriousness to grant giving bodies… in a diverse team where there are a lot of clinicians and basic scientists, it's quite difficult trying to get the mix right as to who decides which direction we go in and how we do that. And the secret is to encourage people and to develop the institutional ethos of research. How do you encourage people to be honest and open about what they do, the mistakes and so forth? How do you develop risk taking safely? What does that mean in the scientific world, what does that mean in the clinical world? They're quite different, but you need a framework. Surgeon 32.

Here discussions on risk are seen as having both similar but largely different sets of meanings in the social worlds of medicine and science. Organ transplantation, and more recently cell transplants, is one area of biomedical science where there has been a long-history of scientists and clinicians working closely together (Fox and Swazey 2002; Fox 1998). Nevertheless, there is a tension between routine clinical care, novel science, and innovative clinical science which itself renders such research collaborations risky. This riskiness is redoubled when placed in the context of perceived resistance to the funding innovative cell transplants:

> [Cell transplantation is] really technical and experimental and so it's falling in to kind of translational research, and not many people are interested in giving money to that area. People either give money to stem cell science but not something like this because it's in between, it's not a clinical practice exactly, and not a basic science either. Physician 34.

The difficulties of working in the 'translational spaces' between medicine and science was universally seen within our wider empirical research as a major barrier to the development of translational stem cell research (see Wainwright et al. 2006b). One reason for this is the perceived increased demands of performing even one of these roles, let alone both:

> We are losing the skills of integration. It's virtually impossible to be a basic scientist and a clinician. You can't do both. You see doctors coming through now who say, 'I can't be bothered, why do I need to do research?' It's really, really sad. I love research. I love clinical research. It's the thing that makes life worthwhile and clinical practice worthwhile. Physician 37.

Inevitably, scientists were seen as lacking in the expertise to make clinical judgements about, say, immunosuppression to prevent rejection of cell transplants:

> My scientist is very good, but he is a scientist, he's not a doctor. And so without the collaboration with the liver team I would not have had the confidence to say, 'We can take these [cells] and put them into a patient the same as I would any product out of the pharmacy.' Physician 31.

In this example, the physician is discussing the link between their small islet cell transplant programme (for diabetes) and the clinicians in the very large liver transplant programme. So what are we witnessing when we analyse some elements of the different fields of medicine and science, and how does this bear on our broad theme of the shaping of 'risky positions'? One possible answer is that risks are shared within a group of scientific and clinical experts who discuss and minimise risk when they work well together in teams:

> Probably the secret is, again, a complex team working and particularly the close collaboration between the scientists and the clinical team. I think that's the environment which is most likely to produce success. All members of the team are contributing and have an input into what actually happens. Surgeon 32.

What we see in these accounts, over and above the external risks posed by poor funding opportunities, is an enactment of risks internal to the process of multidisciplinary collaboration. The sheer difficulty of making this collaboration work is evoked through such vague notions as 'frameworks' (whose?) and 'closeness' (in whose terms?). Performatively, such accounts ground the risks of failure not only in the resistance of 'nature' (e.g. amassing stem cells) but also in the recalcitrance of social relations (e.g. funding, the ever-present possibility of disciplinary fragmentation).

5.3   Experimental Treatments

In this section we discuss the contrast between accounts of life saving and of life-enhancing experimental treatments. Despite what many see as the undermining of trust in modern British medicine, brought about through various controversies such as 'the Bristol and Alder Hay Scandals', there is still more support for 'risky experimental work' that occurs in the domain of medicine than in the domain of science. This is tied up to the making of bold decisions.

> Who took it [transplantation] forward? It was the clinicians. People had to bold about it. Surgeon 46.

As noted above, in fields like organ and cell transplants it was doctors who embarked on 'experiments perilous' (Fox 1998) as they had 'the courage to fail' (Fox and Swazey 2002). But this riskiness is also a reflection of needs of publics, especially those faced with severe medical problems:

> In some respects it's easier when you're dealing with patients than when we started out. But your decisions technically are often black or white in that, if you're going to die and you have an experimental treatment, most patients will take it and ethically one feels comfortable offering it, providing that one explains clearly what the risks are and the background leading up to making that decision. Surgeon 32.

But there is a danger of slippage here. On the one hand, the risks posed by such experimental clinical interventions are not always transparent:

> The general population doesn't understand the difference between clinical and research. I don't think we explain it very well as scientists. And I think we're all too happy to allow people to think that research is something in a nasty dark laboratory with people doing scary things, where anything clinical is safe. And neither view is correct. It's somewhere in between the two. Physician 37.

Risky interventions are also a response to the demands of patients:

> You could put highly speculative trials for cancer patients and get huge numbers of volunteers in what would be ethically unacceptable, just simply because people are desperate and will take any potential treatment, no matter how unrealistic it may be. Funnily enough I think there would be this impedance to carrying out unethical studies, not from the public, it's from the scientists and doctors, who do have ethical standards, which don't actually necessarily match up to the public. You know, we see ourselves as protectors of the public, and I'm not really sure the public sees us the same way… I think it underestimates how willing the public are to experiment, although the problem is, when the experiments don't go right, and that's when the judgement is questioned. Surgeon 32.

A physician who specialises in diabetes evocatively highlights the pressure from patients for experimental treatment:

> They have to think about what they eat every time they eat. You live around your diabetes. And to say to someone like that, 'We could get rid of this for you', they would sell their souls for that. Physician 31.

This ethos of making bold decisions about potentially life saving treatments can thus be contrasted to the risks posed by the demands of patients. Here, we see a dilemma enacted of 'underselling cell transplants' to the public, and particularly patients; and 'overselling cell transplants' to the media, and especially research funding bodies. Exemplifying the latter is the following quote from another one of our interviews with a physician:

> In my opinion it's [cures via cell transplants] a long way away… And you can't get a paper published now unless you make some outrageous statement about its value to the greater community. And a lot of that comes from the pressure of [research funding]. You need an exciting opening sentence. It's like writing a novel. And often it is writing a novel, let's face it! Physician 31.

In sum, what we have seen in this section is the grappling with divergent risks derived respectively from clinical ethos and patient demand. Performatively, this enacts the 'mutual contradictoriness' of medicine and publics. Clinicians must at one make risky decisions and ensure ethical and safe conduct around patients; publics demand both bold clinical intervention, and ethically sound judgement and safe medical practices. In this account, clinicians seem to be caught between the rock of courage and the hard place of caution.

5.4   Regulation and Professional Autonomy

In this section, we investigate some of the ways that regulation has come to affect the ethical 'risky positions' that scientists and doctors engaged in stem cell translation can enact. All our participants remarked upon the necessity of far stricter regulation governing both science and medicine in the UK. Yet, at the same time, this shift in 'discourses of risk' was also seen as problematic. For example, regulatory conditions were seen to have limited the seeming autonomy of transplant clinicians:

> A lot of the research and clinical drive that was there ten, twenty, thirty years ago, really drove clinical medicine forward. It's now not allowed. Liver transplantation wouldn't have happened if it was happening now, because the mortality was too high – Starzl wouldn't be allowed to do it. Physician 37.

Further, the amount of time that is now required for 'formal ethical deliberation' in medicine was seen to hamper the 'professional' efforts of clinicians:

> I love clinical research. It's the thing that makes life worthwhile and clinical practice worthwhile. But yes equally you get so drummed by, 'oh you've got to do the ethics first, you've got to do the audit, you've got to do the morbidity, mortality, the governance, the this, the that,' how many hours are there in a day? Physician 37.

Moreover, the ethics of 'regulatory ethical deliberation' was questioned, not least because such deliberation might itself generate risks (of failing to progress clinically):

I think if you go back 15 or 20 years, we made all these decisions informally as a group of clinicians and drove it on. And whether they were good ethical or bad ethical decisions, was more a reflection of the institutional standing. Now it's become a little more formalised in that if we want to do something the first time, there is a novel treatments committee where one puts together the data where you've got to do your research and you ask permission to do it. Is it a more ethical approach? I'm not really sure, because often these committees don't have good insight, don't have a good feeling for the reality of what you're trying to do. When I go to see a patient and say, 'What you should and shouldn't do,' depends on my honesty. But how I present the facts may be perceived by me as honest, but may be a nuance is placed in certain parts of the speech to give a different meaning. If you wrote it down on the page, it might have a very different impact from how I said it to the patient. Surgeon 32.

When I first came here in '88, we were still consenting people for transplant and it was approaching 50% mortality. You know, NICE wouldn't have allowed it! It would have said it wasn't cost effective. And that's a bit of a tragedy. We are so evidence based driven without really understanding how you generate evidence. And the nature of things is one thing never cured anything. It is a composite. And we've just become a bit too paper driven and evidence based. Physician 37.

Moreover, to the extent that such regulatory ethical deliberation is concerned with the minimization of risk, it was argued that this could also diminish the prospects of clinical learning, prospects that derive from the very possibility of failure:

When I look back, it's nice to see successful transplants. Mostly I think I'm more motivated by failures, like why did it go wrong, where have we got it wrong, how can we do it in a different way to improve it? Surgeon 32.

In the preceding quotes, we see the enactment of another risk dilemma faced by clinicians – that between the necessity for regulatory ethics to assess risks and the constraints such ethical deliberation imposes. This dilemma is placed ambiguously in relation to the risks faced but clinical researchers in other parts of the world:

In India and China, for example, where it's a lot easier to do the research, even though people might criticise them for being unethical and things like that, those were the things that were being done here and those were things that couldn't be done now, because of regulations more than anything else. And that's not fair. That sort of thing is not good. Surgeon 46.

Here, our respondent is bemoaning the broader risks posed by the regulation of risks – risks concerning intellectual, and presumably economic, competition. What is ambiguous in this quote is whether the regulatory regimes at 'home' should be eased, or those 'abroad' tightened. There are of course ironies here, not least given the extremely liberal regulation on the 'in principle' use of human stem cells in the UK. Be that as it may, the point is that such enactments of risk dilemmas, also serve in the 'making' of seemingly distant actors such as 'India and China' and thus, the partial 'making' of those parts of the regulatory system concerned with competition. Put baldly, this rhetorical move runs as follows: if you want us to compete globally, let us have regulatory regimes that favour innovation at home. In this final example, we witness the complexity of the performativity of risk, and how, indeed, risk is a performative resource in the 'making' of various constituencies.

6   Concluding Remarks

In this paper we presented a review of the problems and prospects of stem cell biology and regenerative medicine, most especially in regard of liver disease. We have considered these problems and prospects in terms of clinicians' accounting for the risks associated with, for example, certain emergent medical techniques, the pursuit of particular research programmes, the nature of interdisciplinary collaboration, the utility of clinical 'courage', the pressures imposed by public and patients, the constraints and enablements afforded by regulatory ethical regimes. We have treated these accounts of risk as enactments that performatively 'make' clinicians themselves, but also various other 'constituencies' – notably, public and regulatory actors, though also such diverse 'entities' as scientific rigour, ethics, and other countries. Along the way, we have seen how such accounting for the risks of clinical research convolute to encompass the interplay of multiple risks, not least that between medical risks and social 'meta-risks'.

More specifically, we have begun to trace how clinicians tend to represent themselves as (and this is our heuristic phrase) caught between the 'rock of courage' and the 'hard place of caution'. A common discursive motif, one that is evidently grounded in the perceived history of transplant surgery, has been that of having the 'courage to fail', of taking risks with 'experiments perilous'. Such courage is not just a matter of the character of 'group/grid' configuration or 'habitus' (e.g. Bourdieu 1984) of transplant clinicians – it is enacted in relation to the ostensible expectations of other constituencies, not least that of publics and patients. As we have argued, such boldness is seen to be mediated by a whole range of factors which we might call 'resource constraints'. These limited resources range across the shared risk perceptions with collaborating scientists, 'ethical autonomy', and the accumulation of funds, clinical expertise and materials. Yet, to do bold clinical research in the face of such constraints always risks the possibility of transforming courage into recklessness. Conversely, to occupy the 'hard place of caution' generates other risks, notably the lack of clinical progress but also, the failure to address the demands of patient constituencies or maintain intellectual (and economic) standing in the context of international biomedical competition.

In conclusion, these are tentative comments on the nature of risk enactment amongst clinicians, and we would be unwilling to generalise too much from what is, after all, a small sample. However, we do hope that the present performative perspective holds promise for wider analysis of risk in relation to the health field, notably to the burgeoning area of translational research. Most especially, treating 'risk' as a resource that can be deployed in the enactment of social (and indeed material) relations can usefully move us away from seeing risk in terms of, say, technical calculation, or epochal structural change (e.g. Beck 1992). In this way, we can better grasp how actors in medical and biomedical fields – such as our clinicians – occupy a social world in which 'risk' is common currency but can also be deployed, that is, performed, in ways which help, make, unmake and remake that world.

Acknowledgements This paper is based on two research projects: ESRC Stem Cell Initiative RES-340-25-0003 and RES-350-27-0001.

References

Alison, M.R., R. Poulsom, R. Jeffery, A.P. Dhillon, A. Quaglia, J. Jacob, et al. 2000. Hepatocytes from Non-hepatic Stem Cells. *Nature* 406: 257.

Alison, M.R., M.J. Lovell, N.C. Direkze, N.A. Wright, and R. Poulsom. 2006. Stem Cell Plasticity and Tumour Formation. *European Journal of Cancer* 42 (9): 1247–1256.

Beck, U. 1992. *The Risk Society*. London: Sage.

Blau, H.M., T.R. Brazelton, and J.M. Weimann. 2001. The Evolving Concept of a Stem Cell: Entity or Function. *Cell* 105: 829–841.

Bourdieu, P. 1984. *Distinction: A Social Critique of the Judgement of Taste*. London: Routledge.

Brown, N., and M. Michael. 2002. From Authority to Authenticity: The Changing Governance of Biotechnology. *Health, Risk & Society* 4: 259–272.

Burns, C.J., S.J. Persaud, and P.M. Jones. 2004. Stem Cell Therapy for Diabetes: Do We Need to Make Beta Cells? *Endocrinology* 183 (3): 437–443.

Calne, R.Y. 1998. *The Ultimate Gift: The Story of Britain's Premier Transplant Surgeon*. London: Headline.

———. 2003. The History of Liver Transplantation. In *History of Organ and Cell Transplantation*, ed. N.S. Hakim and V.E. Papalois, 100–119. London: Imperial College Press.

Calne, R.Y., and R. Williams. 1968. Liver Transplantation in Man- I. Observations on Technique and Organisation in 5 Cases. *British Medical Journal* 280: 535–540.

Cameron, J.S. 2002. *History of the Treatment of Renal Failure by Dialysis*. Oxford: Oxford University Press.

Clarke, A.E., J.K. Shim, L. Mamo, J.R. Fosket, and J.R. Fishman. 2003. Biomedicalization: Technoscientific Transformations of Health, Illness, and US Biomedicine. *American Sociological Review* 68: 161–194.

Dahlke, M.H., F.C. Popp, S. Larsen, H.J. Schlitt, and J.E.J. Rasko. 2004. Stem Cell Therapy of the Liver – Fusion or Fiction? *Liver Transplantation* 10: 471–479.

Department of Health. 2005. UK Stem Cell Initiative: [Pattison] Report & Recommendations.

Donovan, P.J., and J. Gearhart. 2001. The End of the Beginning for Pluripotent Stem Cells. *Nature* 414: 92–97.

Douglas, M. 1970. *Natural Symbols: Explorations in Cosmology*. New York: Pantheon.

Evans, M.J., and M.H. Kaufman. 1981. Establishment in Culture of Pluripotent Stem Cells from Mouse Embryos. *Nature* 292: 154–156.

Fausto, N. 2004. Liver Regeneration and Repair: Hepatocytes, Progenitor Cells, and Stem Cells. *Hepatology* 39: 1477–1487.

Forsythe, J.L.R., ed. 2001. *Transplantation Surgery: Current Dilemmas*. Philadelphia: Saunders.

Fox, R.C. 1998. *Experiment Perilous Physicians and Patients Facing the Unknown*. New Brunswick: Transaction.

Fox, R.C., and J.P. Swazey. 2002. *The Courage to Fail: A Social View of Organ Transplants and Dialysis*. 3rd ed. New Brunswick: Transaction.

Franklin, S. 2001. Culturing Biology: Cell Lines for the Second Millennium. *Health* 5: 335–354.

———. 2005. Stem Cells R Us: Emergent Life Forms and the Global Biological. In *Global Assemblages: Technology, Politics and Ethics as Anthropological Problems*, ed. A. Ong and S.J. Collier. New York: Blackwell.

Gearhart, J. 1998. New Potential for Human Embryonic Stem Cells. *Science* 282: 1061–1062.

Gearhart, J., ed. 2005. *Stem Cells: Nuclear Reprogramming and Therapeutic Applications*. *Novartis Foundation Symposium*. New York: Wiley.

Grompe, M. 2004. The Importance of Knowing Your Identity: Sources of Confusion in Stem Cell Biology. *Hepatology* 39: 35–37.

Hacking, I. 1986. Making Up People. In *Reconstructing individualism*, ed. T.C. Heller, M. Sosna, and D.E. Wellberg. Stanford: Stanford University Press.

Hamilton, C., S. Adolphs, and B. Nerlich. 2007. The Meanings of 'Risk': A View from Corpus Linguistics. *Discourse & Society* 18: 163–181.

Jasanoff, S. 2005. *Designs on Nature: Science and Democracy in Europe and the United States*. Princeton: Princeton University Press.

Johnson, B.B. 1987. The Environmentalist Movement and Grid/Group Analysis: A Modest Critique. In *The Social Construction of Risk*, ed. B.V. Covello and B. Johnson. Dordrecht: Reidel.

Keating, P., and A. Cambrosio. 2003. *Biomedical Platforms: Realigning the Normal and the Pathological in Late-Twentieth Century Medicine*. Cambridge: MIT Press.

Kitzinger, J., and C. Williams. 2005. Forecasting Science Futures: Legitimising Hope and Calming Fears in the Embryo Stem Cell Debate. *Social Science & Medicine* 61: 731–740.

Lanza, R., H. Blau, J. Gearhart, B. Hogan, D. Melton, M. Moore, et al., eds. 2004. *Handbook of Stem Cells, Volume 1 Embryonic Stem Cells*. Amsterdam: Elsevier.

Lock, M. 2001. *Twice Dead: Organ Transplants and the Reinvention of Death*. Berkeley: University of California Press.

Lupton, D. 1999. *Risk*. London: Routledge.

Michael, M., and N. Brown. 2005. Scientific Citizenships: Self-Representations of Xenotransplantation's Publics. *Science as Culture* 14 (1): 38–57.

Michael, M., S.P. Wainwright, C. Williams, B. Farsides, and A. Cribb. 2007a. From Core Set to Assemblage: On the Dynamics of Exclusion and Inclusion in the Failure to Derive Beta Cells from Embryonic Stem Cells. *Science Studies* 20: 5–25.

Michael, M., S.P. Wainwright, and C. Williams. 2007b. Temporality and Prudence: On Stem Cells as 'Phronesic Things'. *Configurations* 13: 373–394.

Mitry, R., and A. Dhawan. 2002. Hepatocyte Transplantation from Bench to Bedside. *British Inherited Metabolic Disease Group Bulletin*, Autumn, 1–4.

Mol, A. 2002. *The Body Multiple. Ontology in Medical Practice*. Durham: Duke University Press.

Nightingale, P., and P. Martin. 2004. The Myth of the Biotech Revolution. *Trends in Biotechnology* 22: 563–569.

Norman, D.J., and L.A. Turka. 2001. *Primer on Transplantation: Manual of the American Society of Transplantation*. 2nd ed. Oxford: Blackwell Science.

O'Grady, J., J. Lake, and P. Howdel. 2006. *Comprehensive Clinical Hepatology*. 2nd ed. St Louis: Mosby.

Parry, S. 2003. The Politics of Cloning: Mapping the Rhetorical Convergence of Embryos and Stem Cells in Parliamentary Debates. *New Genetics & Society* 22: 145–168.

Plevris, J.N., and P.C. Hayes. 2001. Bioengineering as an Alternative to Liver Transplantation. In *Transplantation Surgery: Current Dilemmas*, ed. J.L.R. Forsythe. Philadelphia: Saunders.

Pickering, S.J., P.R. Braude, M. Patel, C.J. Burns, J. Trussler, V. Bolton, and S. Minger. 2003. Preimplantation Genetic Diagnosis as a Novel Source of Embryos for Stem Cell Research. *Reproductive Biomedicine Online* 7 (3): 353–364.

Polak, J.M., L.L. Hench, and P. Kemp. 2002. *Future Strategies for Tissue and Organ Replacement*. London: Imperial College Press.

Rajan, K.S. 2006. *Biocapital: The Constitution of Postgenomic Life*. Durham: Duke University Press.

Rayner, S. 1986. Management of Radiation Hazards in Hospitals: Plural Rationalities in a Single Institution. *Social Studies of Science* 16: 573–591.

Rose, N. 2007. *The Politics of Life Itself*. Princeton: Princeton University Press.

Sargent, S., and S.P. Wainwright. 2006. Quality of Life Following Emergency Liver Transplantation for Acute Liver Failure. *Nursing in Critical Care* 11: 168–176.

Schwarz, M., and M. Thompson. 1990. *Divided We Stand*. Hemel Hempstead: Harvester Wheatsheaf.

Scott, C.T. 2006. *Stem Cell Now: From the Experiment That Shook the World to the New Politics of Life*. New York: Pi Press.

Seldon, C., and H. Hodgson. 2002. Engineering the Liver. In *Future Strategies for Tissue and Organ Replacement*, ed. J.M. Polak, L.L. Hench, and P. Kemp. London: Imperial College Press.

Starzl, T.E. 2003. *The Puzzle People: Memoirs of a Transplant Surgeon*. Pittsburgh: University of Pittsburgh Press.

Starzl, T.E., T.L. Marchioro, K.N. Von Kaulla, G. Hermann, R.S. Brittain, and W.R. Waddell. 1963. Homotransplantation of the Liver in Humans. *Surgery Gynecology & Obstetrics* 117: 659–665.

Strauss, A.L. 1987. *Qualitative Analysis for Social Scientists*. Cambridge: Cambridge University Press.

Suknikh, G.T., and A. Shitl. 2003. Stem Cell Transplantation for Treatment of Liver Diseases. *International Journal of Molecular Medicine* 11: 395–400.

Theise, N. 2003. Liver Stem Cells: The Rise and Fall of Tissue Biology. *Hepatology* 38: 804–806.

Thomson, J.A., J. Itskovitz-Eldor, S.S. Shapiro, M.A. Waknitz, J.J. Swiergiel, V.S. Marshall, et al. 1998. Embryonic Stem Cell Lines Derived from Human Blastocysts. *Science* 282: 1145–1147.

Tulloch, J., and D. Lupton. 2003. *Risk and Everyday Life*. London: Sage.

Wainwright, S.P. 1994. Recovery from Liver Transplantation: A Literature Review. *Journal of Advanced Nursing* 20: 861–869.

———. 1995. The Transformational Experience of Liver Transplantation. *Journal of Advanced Nursing* 22: 1068–1076.

———. 1997. Transcending Chronic Liver Disease: A Qualitative Study. *Journal of Clinical Nursing* 6: 43–53.

Wainwright, S.P., and D. Gould. 1997. Non-adherence with Medications in Organ Transplant Patients: A Literature Review. *Journal of Advanced Nursing* 26: 968–977.

Wainwright, S.P., C. Williams, M. Michael, B. Farsides, and A. Cribb. 2006a. From Bench to Bedside? Biomedical Scientists' Expectations of Stem Cell Science as a Future Therapy for Diabetes. *Social Science & Medicine* 63: 2052–2064.

———. 2006b. Ethical Boundary Work in the Embryonic Stem Cell Laboratory. *Sociology of Health & Illness* 28: 732–748.

Wainwright, S.P., C. Williams, S.J. Persaud, and P.M. Jones. 2006c. Real Science, Biological Bodies and Stem Cells: Constructing Images of Beta Cells in the Biomedical Science Lab. *Social Theory & Health* 4: 275–298.

Wainwright, S., C. Williams, M. Michael, B. Farsides, and A. Cribb. 2006d. Ethical Boundary Work in the Embryonic Stem Cell Laboratory. *Sociology of Health & Illness* 28: 732–748.

Wainwright, S.P., C. Williams, M. Michael, B. Farsides, and A. Cribb. 2007. Remaking the Body? Scientists' Genetic Discourses and Practices as Examples of Changing Expectations on Embryonic Stem Cell Therapy for Diabetes. *New Genetics & Society* 26: 251–268.

Waldby, C. 2002. Stem Cells, Tissue Cultures and the Production of Biovalue. *Health* 6: 305–323.

Weber, R. 1990. *Basic Content Analysis*. London: Sage.

Williams, C., J. Kitzinger, and L. Henderson. 2003. Envisaging the Embryo in Stem Cell Research: Discursive Strategies and Media Reporting of the Ethical Debates. *Sociology of Health & Illness* 25: 793–814.

Williams, C., S.P. Wainwright, K. Ehrich, and M. Michael. 2008. Human Embryos as Boundary Objects? Some Reflections on the Biomedical Worlds of Embryonic Stem Cells and Pre-implantation Genetic Diagnosis. *New Genetics & Society* 27: 7–18.

Outroduction

Hauke Riesch, Nathan Emmerich, and Steven Wainwright

In the introduction to this volume we have argued that being an interdisciplinary scholar involves managing a complex interplay of disciplinary identities, as well as the ontologies and ways of knowing and understanding that are associated with the subject matter(s). We argued that trying to force a bioethical interdiscipline without a special regard to the individual epistemological, ontological and social aspects of the disciplines is unlikely to bear fruit in the long-term. Although bioethics has always been a multidisciplinary activity, the relations between the various disciplines involved have traditionally been fraught, particularly so at the intersection between sociological and philosophical bioethics. We therefore proposed approaching bioethical interdisciplinarity in a manner that is careful and measured. This does not involve abandoning disciplines or disciplinarity, rather we are advocating for an increased level of cross-fertilisation between established disciplines or the way in which research is done, in the broadest sense.

Setting up cross-disciplinary fora such as an edited volume that includes contributions from both sides may be the best way forward in the first instance. However, though we believe that generalised and overly normative prescriptions on how one should "do" interdisciplinarity are rarely helpful, we might attend to the contributions to this volume in order to reach some conclusions as to how work on and across the philosophy/sociology boundary can be done, particularly in the case of bioethics.

H. Riesch (✉) · S. Wainwright
Department of Social and Political Sciences, Brunel University London, London, UK
e-mail: Hauke.riesch@brunel.ac.uk; Steven.wainwright@brunel.ac.uk

N. Emmerich
ANU Medical School, Australian National University, Canberra, Australia

The Institute of Ethics, Dublin City University, Dublin, Ireland

School of History, Anthropology, Politics and Philosophy,
Queen's University Belfast, Belfast, UK
e-mail: nathan.emmerich@anu.edu.au

© Springer International Publishing AG, part of Springer Nature 2018   171
H. Riesch et al. (eds.), *Philosophies and Sociologies of Bioethics*,
https://doi.org/10.1007/978-3-319-92738-1_10

Although it should go without saying, we might first note that interdisciplinary interactions will not be fruitful unless they are based on mutual respect. When arguments from the perspective of one discipline call into question the central purpose of the other, or when one discipline intimates that it should supplant the other, suggesting that all work the latter discipline has traditionally done should now fall under the auspices of former, then the attempt at interdisciplinarity has failed at the first hurdle.[1] Thus, whilst there are arguments to be made about whether bioethics research *should,* say, be normative or *should* be descriptive, we think that there is space enough for both as long as they're not done in isolation.

Beyond the very obvious need for respect between the disciplines, and that bringing contributors together in a shared forum such as this begins to address interdisciplinarity, the contributions to our collection make further inroads into 'interdisciplinary bioethics.' This often results from the particular strategies adopted within each chapter. We would like to conclude by considering these strategies and evaluating if and how they match up with our stated purpose of "crossing the divides." In short, we would like to directly consider whether or not we have collectively been successful in producing the kind of work we sought to engage in and promote, at least as a first step in the broader project of greater and better interdisciplinary cooperation between sociology and philosophy.

1. Sociological work as case studies grounding philosophical discussion: Riesch, Paton, and Wainwright et al. have all offered sociological work that opens up questions that philosophers are working on.
2. Emmerich, Rappert and Kendig have taken theoretical foundations of a larger disciplinary background – Science and Technology Studies and History and Philosophy of Science – and applied these to bioethics issues. Interestingly, both these larger disciplines are themselves already "interdisciplinary" disciplines.
3. Tamimi and del Savio presented a more integrated ethical discussion based on their larger projects on vaping and citizen science.

Given these comments on the chapters presented in this book, it would seem that 'bridging the divides' might mean producing work that acknowledges the relevance to adjacent fields and being forthright in including philosophical bioethical concerns within a largely sociological study. It could also mean taking inspiration from the theoretical groundwork of one discipline, or wider disciplinary grouping, and making use of it in order to present a novel account of a particular bioethical issue or question. Or it could mean integrating philosophical bioethical discussion directly into an empirical framework and following the paths this leads us to. Of course, this list of doing interdisciplinarity is not exhaustive; we could for example imagine philosophical studies consciously preparing the ground for future sociological ones, and vice versa.

---

[1] We should be careful with our condemnations. It can be the case that such arguments can be reasonable and ought to be considered on their merits. However, contributing to these debates was not our intention here.

These individual strategies may be encapsulated through wider terminological divisions that have been introduced in discussions on interdisciplinarity. These include transdisciplinarity, multidisciplinarity, and pluridisciplinarity, with each one presenting one particular way of doing interdisciplinarity. However if there is a larger lesson on how this collection of essays demonstrates a way forward, then it is insofar as it shows that there are multiple ways in which the divide might be bridged. It is in demonstrating these different ways that progress is to be made. Thus, we would suggest that a particular virtue of this collection is that it can point towards future engagement within, between and across philosophical and sociological bio-ethics and, we hope, for academics of other disciplinary identities when it comes to thinking about interdisciplinarity in the domain of bioethics.

The manufacturer's authorised representative in the EU is Springer
Nature Customer Service Centre GmbH, Europaplatz 3, 69115 Heidelberg,
Germany. If you have any concerns regarding our products, please
contact ProductSafety@springernature.com

Printed and bound by CPI Group (UK) Ltd, Croydon, CR0 4YY
23/04/2026
02095634-0001